Mind Performance Projects for the Evil Genius: 19 Brain-Bending Bio Hacks

Evil Genius Series

Bike, Scooter, and Chopper Projects for the Evil Genius

Bionics for the Evil Genius: 25 Build-It-Yourself Projects

Electronic Circuits for the Evil Genius: 57 Lessons with Projects

Electronic Gadgets for the Evil Genius: 28 Build-It-Yourself Projects

Electronic Games for the Evil Genius

Electronic Sensors for the Evil Genius: 54 Electrifying Projects

46 Science Fair Projects for the Evil Genius

50 Awesome Auto Projects for the Evil Genius

50 Green Projects for the Evil Genius

50 Model Rocket Projects for the Evil Genius

51 High-Tech Practical Jokes for the Evil Genius

Fuel Cell Projects for the Evil Genius

Mechatronics for the Evil Genius: 25 Build-It-Yourself Projects

Mind Performance Projects for the Evil Genius: 19 Brain-Bending Bio Hacks

MORE Electronic Gadgets for the Evil Genius: 40 NEW Build-It-Yourself Projects

101 Outer Space Projects for the Evil Genius

101 Spy Gadgets for the Evil Genius

123 PIC® Microcontroller Experiments for the Evil Genius

123 Robotics Experiments for the Evil Genius

125 Physics Projects for the Evil Genius

PC Mods for the Evil Genius

Programming Video Games for the Evil Genius

Solar Energy Projects for the Evil Genius

Telephone Projects for the Evil Genius

22 Radio and Receiver Projects for the Evil Genius

25 Home Automation Projects for the Evil Genius

Mind Performance Projects for the Evil Genius: 19 Brain-Bending Bio Hacks

BRAD GRAHAM and KATHY McGOWAN

New York Chicago San Francisco Lisbon
London Madrid Mexico City Milan New Delhi
San Juan Seoul Singapore Sydney Toronto

The McGraw-Hill Companies

McGraw-Hill books are available at special quantity discounts to use as premiums and sales promotions, or for use in corporate training programs. To contact a representative, please e-mail us at bulksales@mcgraw-hill.com.

Mind Performance Projects for the Evil Genius: 19 Brain-Bending Bio Hacks

1 2 3 4 5 6 7 8 9 0 QPD/QPD 0 1 5 4 3 2 1 0 9

ISBN 978-0-07-162392-6
MHID 0-07-162392-2

Sponsoring Editor
Judy Bass

Editing Supervisor
Stephen M. Smith

Production Supervisor
Richard C. Ruzycka

Acquisitions Coordinator
Michael Mulcahy

Project Manager
Patricia Wallenburg, TypeWriting

Copy Editor
James K. Madru

Proofreader
Paul Tyler

Indexer
Karin Arigoni

Art Director, Cover
Jeff Weeks

Composition
TypeWriting

For Trudy "SilverWolf." Thank you for your wonderful support and friendship.

—NorthernWolf

About the Authors

Brad Graham is an inventor and electronics hobbyist. He is the coauthor of *101 Spy Gadgets for the Evil Genius*, *Atomic Zombie's Bicycle Builder's Bonanza*, and *Bike, Scooter, and Chopper Projects for the Evil Genius*, all from McGraw-Hill Professional.

Kathy McGowan is Mr. Graham's other "evil genius" half, providing administrative, logistical, and marketing support for Atomic Zombie's many human-powered, electric, technical, and publishing projects. She also manages the daily operations of their high-tech firm and numerous Web sites.

Contents

Acknowledgments

Sincerest thanks always to our "evil genius" collaborator, Judy Bass, senior and series editor at McGraw-Hill Professional in New York. Your enthusiasm and support keep us motivated to forge ahead and never give up. You are a very special lady.

Also thanks to many people who take an interest in our projects and encourage us to pursue our dreams. We would especially like to thank Paul Tulonen, industry advisor with the National Research Council of Canada and the Industrial Research Assistance Program, and the extraordinarily talented folks at Digital Engineering, Inc., Thunder Bay, Ontario.

Mind Performance Projects for the Evil Genius: 19 Brain-Bending Bio Hacks

Introduction

Getting Started

Welcome "Noobs"!

If you have been experimenting with electronics for any amount of time, then chances are that you can skip right past this chapter and start digging into some of the projects presented in this book. If you are just starting out, then you have a little groundwork to cover before you begin, but don't worry, the electronics hobby is well within reach for anyone with a creative mind and a desire to learn something new.

As with all things new, you have to start from the beginning and expect a few failures along the way. We electronic "nerds" call this "letting out the magic smoke," and you will fully understand this phrase the first time you connect your power wires in reverse! Please do not be intimidated by the huge amount of technical material available on electronic components and devices because chances are that you need only a small amount of what is available to complete a project. All problems can be broken down into smaller parts, and a schematic diagram is a perfect example of this. Once you understand the basic principles, you will be able to look at a huge schematic or circuit and see that it is made up of smaller basic building blocks just like a brick wall.

Because of limited space in this book, I will cover only the essential basics you need to get started in this fun and rewarding hobby, but there are thousands of resources available to research as you move forward one step at a time. You have the "evil genius" itch; now, all you need is a good pile of junk and a few basic tools to set your ideas into motion!

The Breadboard

This oddly named tool is probably the most important prototyping device you will ever own, and it is absolutely essential to this hobby. A *breadboard* or *solderless breadboard* is a device that lets you connect the leads of semiconductors together without wires so that you can test and modify your circuit easily without soldering. Essentially, it is nothing more than a board full of small holes that interconnect in rows so that you can complete a circuit. In the early days, our "evil genius" forefathers would drive a bunch of nails into an actual board (such as a cutting board for bread) and then connect their components to the nails. So you can thank those pioneering "evil geniuses" who once sat in their workshops with a breadboard full of glowing vacuum tubes and wires for the name!

Today's breadboards look nothing like the originals, often containing hundreds of rows to accommodate the increasing complexity and pin count of today's circuitry. It is common to have 50 or more complex integrated circuits (ICs) on a breadboard running at speeds of up to 100 MHz, so a lot can be done with breadboards. One of my latest breadboard projects was a fully functional 8-bit computer with a double-buffered video graphics array (VGA) output and complex sound generator. This project ran flawlessly on a breadboard at speeds of 40 MHz and had an IC count of over 30, so don't let anyone tell you that a breadboard is only for simple low-speed prototyping. Let's have a look at a typical solderless breadboard that can be purchased at most electronics supply outlets (Figure I-1).

A breadboard such as the one in the figure typically will cost you around $30 and will provide years of use. Without a breadboard, you would have to solder your components together and hope that your design worked on the first try, something that is only a pipe dream in this hobby. The connections under the plastic holes are designed so that the power strips (marked + and –) are connected horizontally, and the prototyping-area holes are connected vertically. The small gutter between the prototyping holes is there so that your ICs can press into the board with each row of legs on each side of the gutter. Figure I-2 shows a close-up of the interconnections underneath the plastic board.

As you can see, the power-strip holes connect horizontally, and the prototyping holes connect vertically. In this way, you can have power along the entire strip because ground (GND) and power (VCC) often have multiple connection points in a circuit. Once you are familiar with a breadboard, it is easy to rig up a test circuit in minutes, even one with a high component count. Once your circuit is tested and working, you can move it to a more permanent home, such as a copper-clad board or even a real printed-circuit board (PCB).

To make the connections from one row of holes to another, you will need wires, many wires. Breadboard wires should be solid, not stranded, have about ¼ in of bare wire at the end, and come in multiple colors and lengths to make tracing your circuit easier. You can purchase various breadboard-ready wiring packs

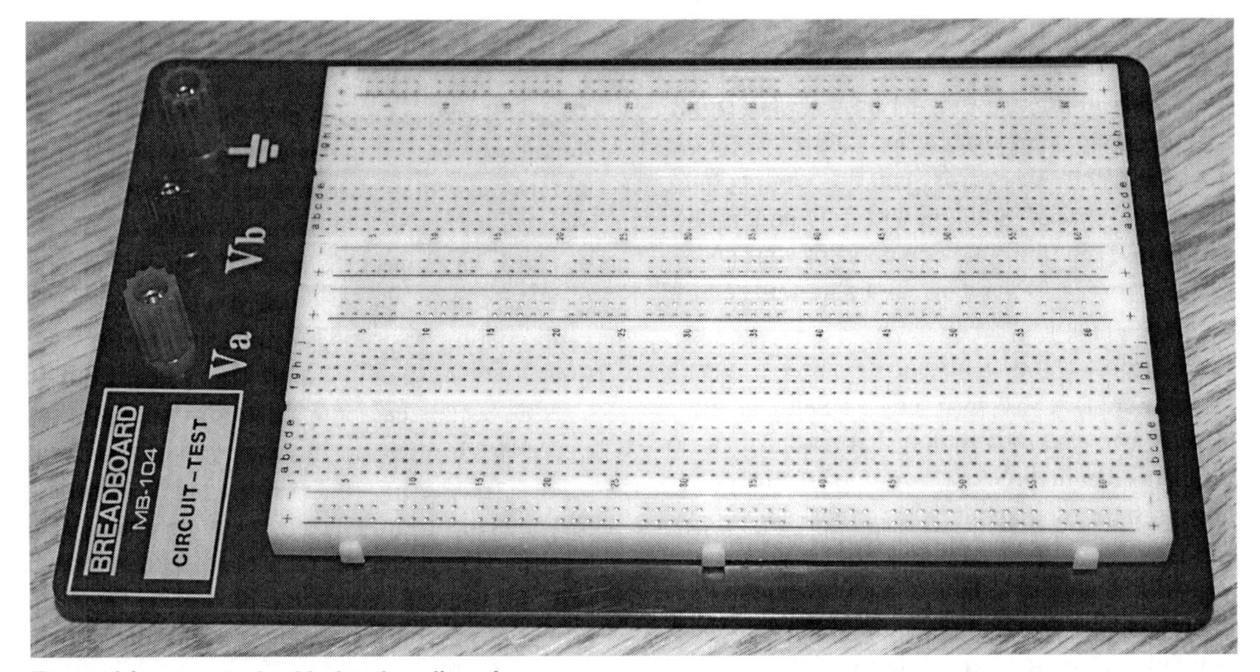

Figure I-1 *A typical solderless breadboard.*

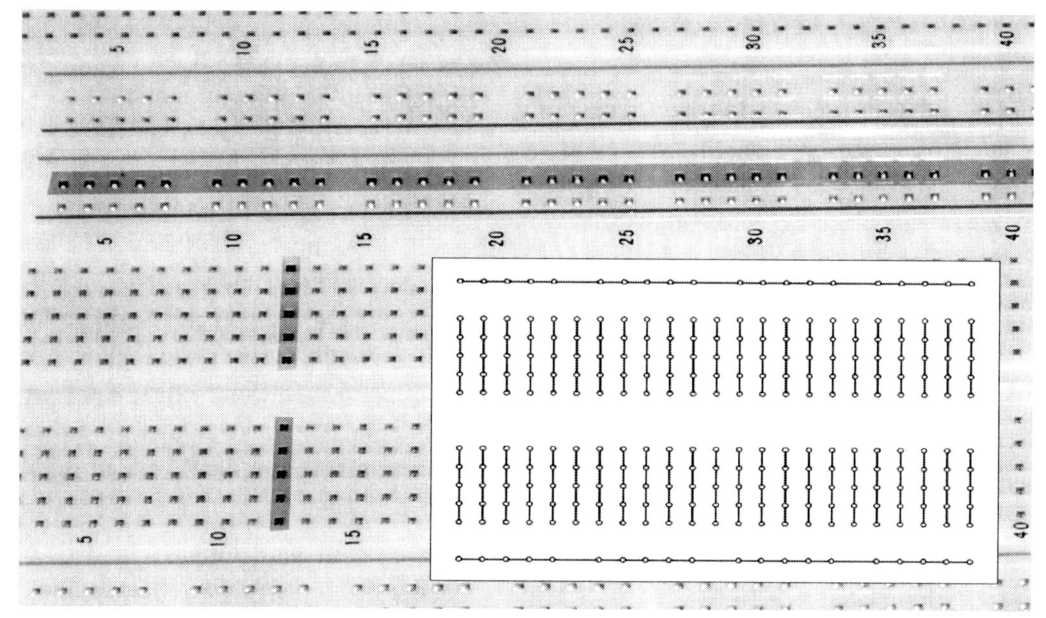

Figure I-2 *Connections between holes on a breadboard.*

from electronics supply stores, but when you get into larger prototyping, it may get expensive to purchase as many wires as you need. The best solution I have found is to get a good length of Category 5 (Cat5) wire, which is used for computer networking, and then strip the ends for use on the breadboard. The nice thing about Cat5 wire is that it has eight colored wires with a solid copper core that are a perfect size to fit into a breadboard. Figure I-3 shows some of the

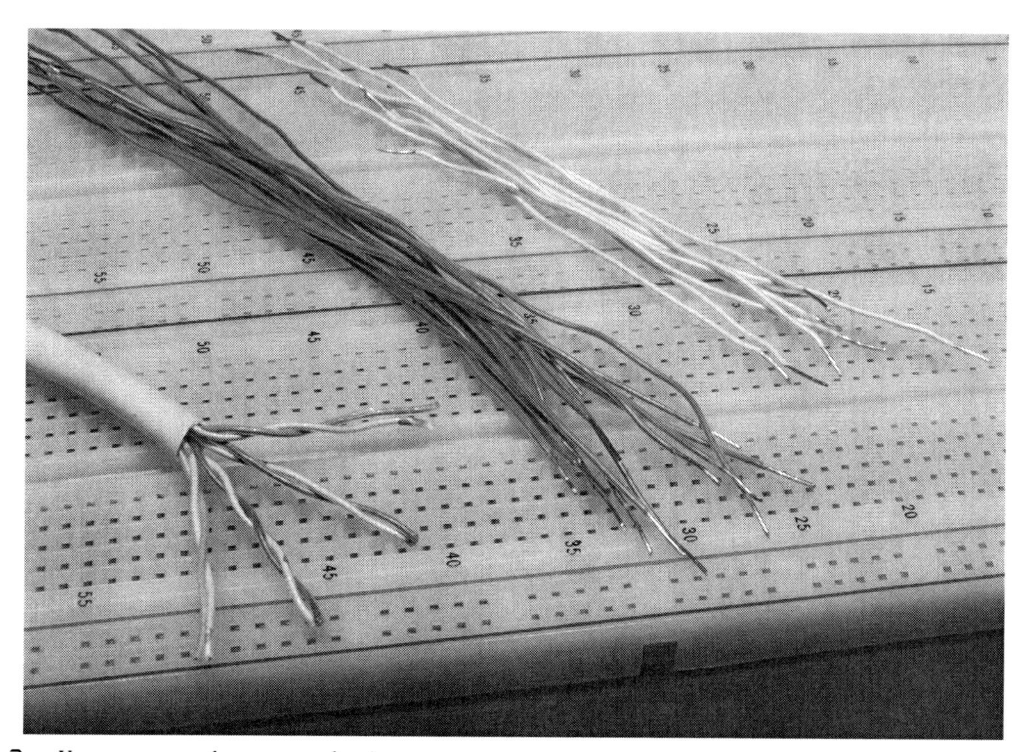

Figure I-3 *You can never have enough wires.*

Cat5 wiring cut and stripped for use in my breadboard.

Cat5 wiring comes as four twisted pairs, so I just cut a bunch of lengths, unwind the wires, and then strip the ends of the plastic sheathing using a dull utility knife. Just place the wire over a dull blade, and press down with your thumb to score the end so that it pulls away from the wire. An actual wire stripper works just as well, but the dull blade seems a lot faster when you want 100 or more wires for your breadboard. For starters, you will need about 20 wires (each) in lengths of 1, 2, 4, and 6 in and a few longer wires for external devices. The 1-in wires also should include red and green (or similar) colors that easily identify your power connections. The power wires will be the most-used wires on your board, so make sure that you have enough of them to go around.

Figure I-4 shows why it is important to have a dull blade for stripping the wires if you choose to do it this way. I purposely sanded the edge of this utility knife so that it would be sharp enough to score the wiring shield yet not cut my thumb after repeated stripping of hundreds of wires in a row. A wire-stripping tool also works well, but I find it to be a little slow when I need 128 blue bus wires for a circuit I want to complete before the end of the night. Now that you have a breadboard and an endless supply of wires to press into the holes, you can begin prototyping your first circuit. Learning to look at a schematic, identifying the semiconductors, and then transplanting the connections to your breadboard will open up an entire world of fun, so let's find out how it is done.

Figure I-5 shows a simple schematic of a capacitor and a resistor wired in parallel. This wiring is transferred to the breadboard by placing the leads of the two components into the holes and then using the wires to join the rows. Since the holes are all connected to each other in vertical rows, the wires can be placed in any one of the five holes that make a row. Although a breadboard circuit may have 500 wires, there is

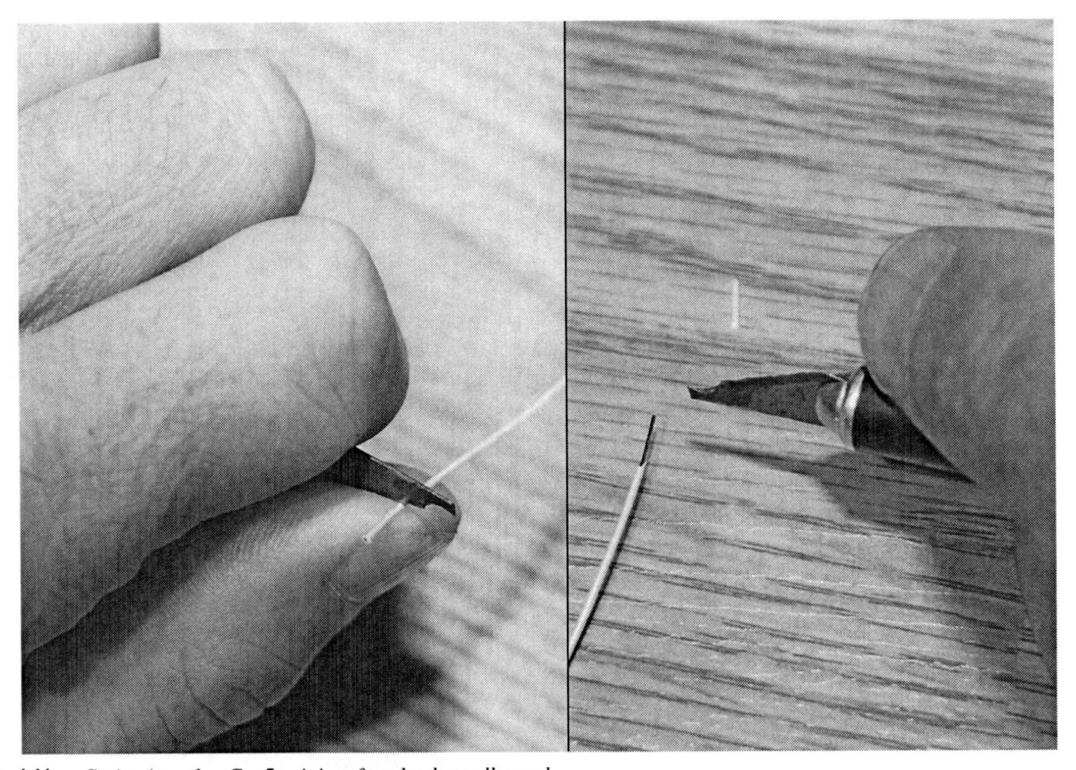

Figure I-4 *Stripping the Cat5 wiring for the breadboard.*

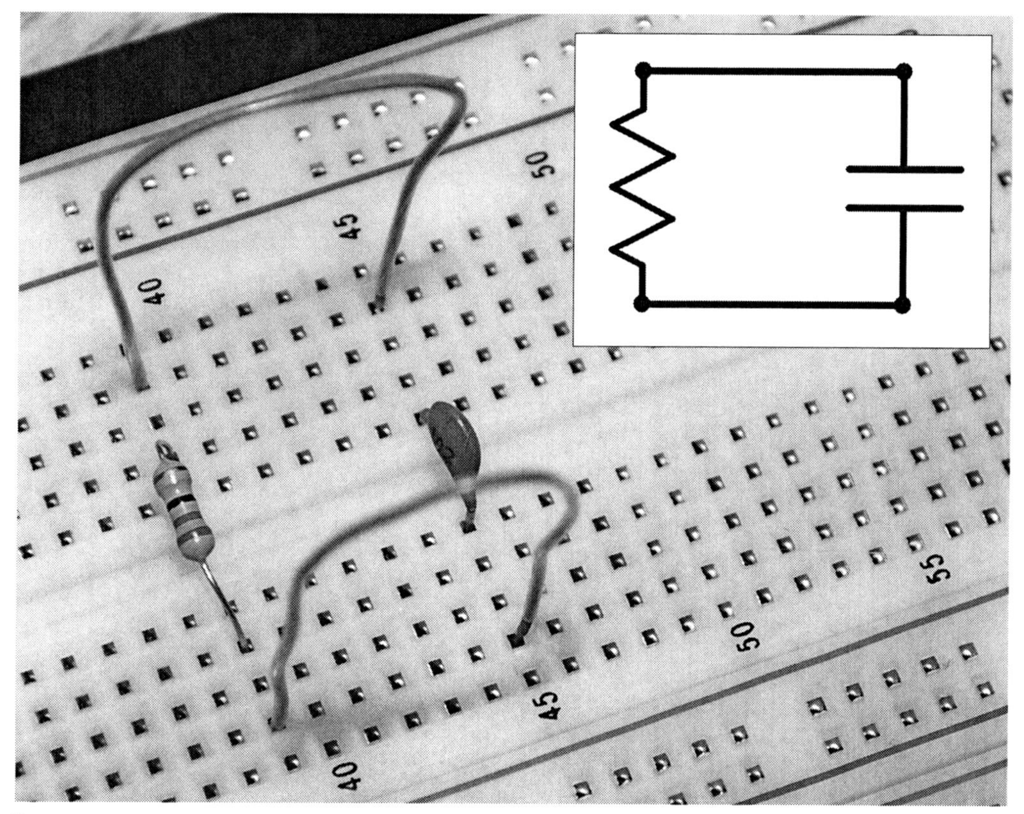

Figure I-5 *Learning to use the breadboard.*

not much more to it than that. Now you can see why prototyping a circuit on a breadboard is easy and lends itself well to modification. There are, however, a few breadboard "gotchas" to keep in mind, and capacitive noise, or *crosstalk*, is one of them.

Crosstalk can be a real problem on a breadboard because the metal plates that make up the rows of holes are close enough together to act as capacitors. This can induce noise in your circuit, possibly causing it to fail or act differently than expected. Radiofrequency (rf) or high-speed digital circuits are prone to noise and crosstalk, and this can create all kinds of "evil genius" headaches. Sometimes you can design a high-speed or rf circuit on a breadboard and have it work perfectly, only to find that it fails completely or acts differently when redone on a permanent circuit board. The capacitance of the plates is actually part of the circuit now! Although you can never really eliminate this error, there is always a way to greatly reduce noise on a breadboard, and it involves adding a few decoupling capacitors on your power strips.

Decoupling capacitors act as filters so that rf and alternating-current (ac) noise don't leak into your power source, causing havoc throughout the entire circuit. When working with high-speed logic, microcontrollers on a clock source, or rf circuits, decoupling capacitors are very important and should not be left out. If you look at an old logic circuit board, you will notice that almost every IC has a ceramic capacitor nearby or directly across the VCC and ground pins. These capacitors are nothing more than 0.01-μF ceramic capacitors placed between VCC and ground on each of your power-supply rails, as shown in Figure I-6. Also notice in the figure that the power rails need to be connected to each other because each strip is independent. If you forget to connect a rail, it will neither carry VCC nor be grounded, so your circuit will fail. Usually,

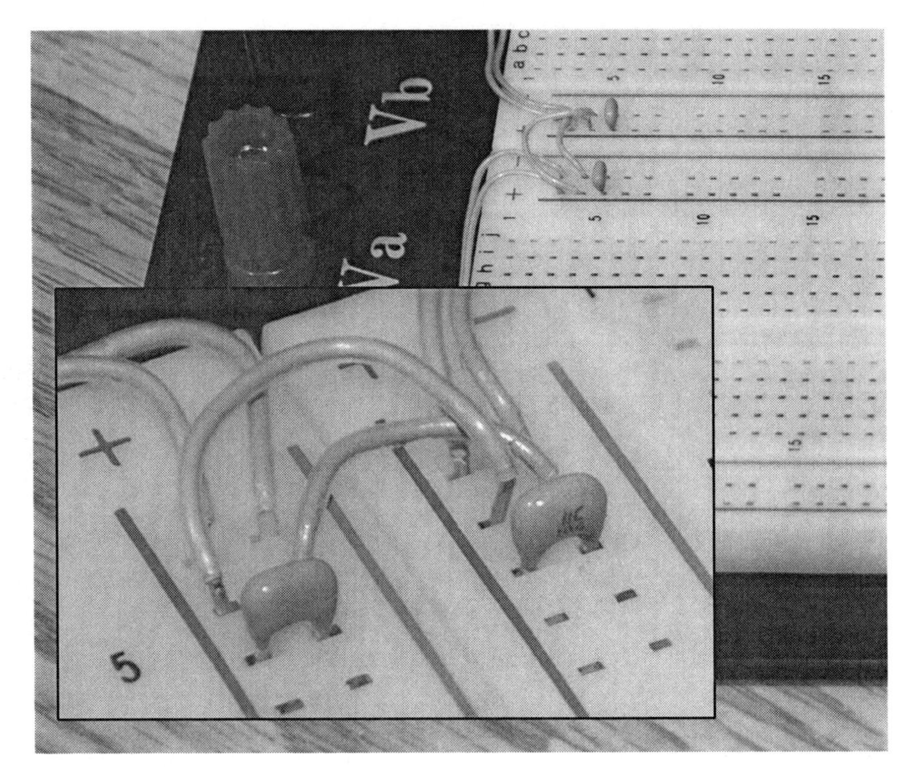

Figure I-6 *Fighting breadboard noise.*

adding decoupling capacitors on each end of the powers strips will be adequate, but in a large high-speed or rf circuit, you may need them closer to the IC power lines or other key components.

When your designs become very large or complex, the standard breadboard may not offer enough real estate, but not to worry, you can purchase individual breadboard sections and snap them together to make a larger prototyping area. Figure I-7 shows 10 breadboard sections snapped together and then bolted to a steel cookie sheet in order to create a grounding plate. The steel base also helps to reduce noise, and all breadboards should have a metal base. This massive cookie-sheet breadboard circuit is a fully functional 20-MHz video computer that can display high-resolution graphics to a VGA monitor and generate complex multichannel sound. The entire computer was designed on the breadboard shown in the figure and went directly to the final design stage based on the breadboard circuit. I have built high-speed computer systems running at more than 75 MHz on a breadboard as well as high-power video transmitters, robot motor controllers, and every single project in this book. Until you break the 100-MHz barrier, there is not much you can't do on a breadboard, so become good friends with this powerful prototyping tool!

Electronic Building Blocks

So you've just found a cool schematic on the Internet, and you have a brand-new breadboard with 100 wires waiting to find a home, but where do you get all those components? If you have been doing this for a while, then your "junk box" is probably well equipped, but for those just starting out, you have to be resourceful to keep your budget under control. A simple circuit with 10 small components might cost you only $5 at the local electronics shop, but often you may need a lot more than 10 parts or might need some uncommon semiconductors. The best source for free electronic components is from old circuit

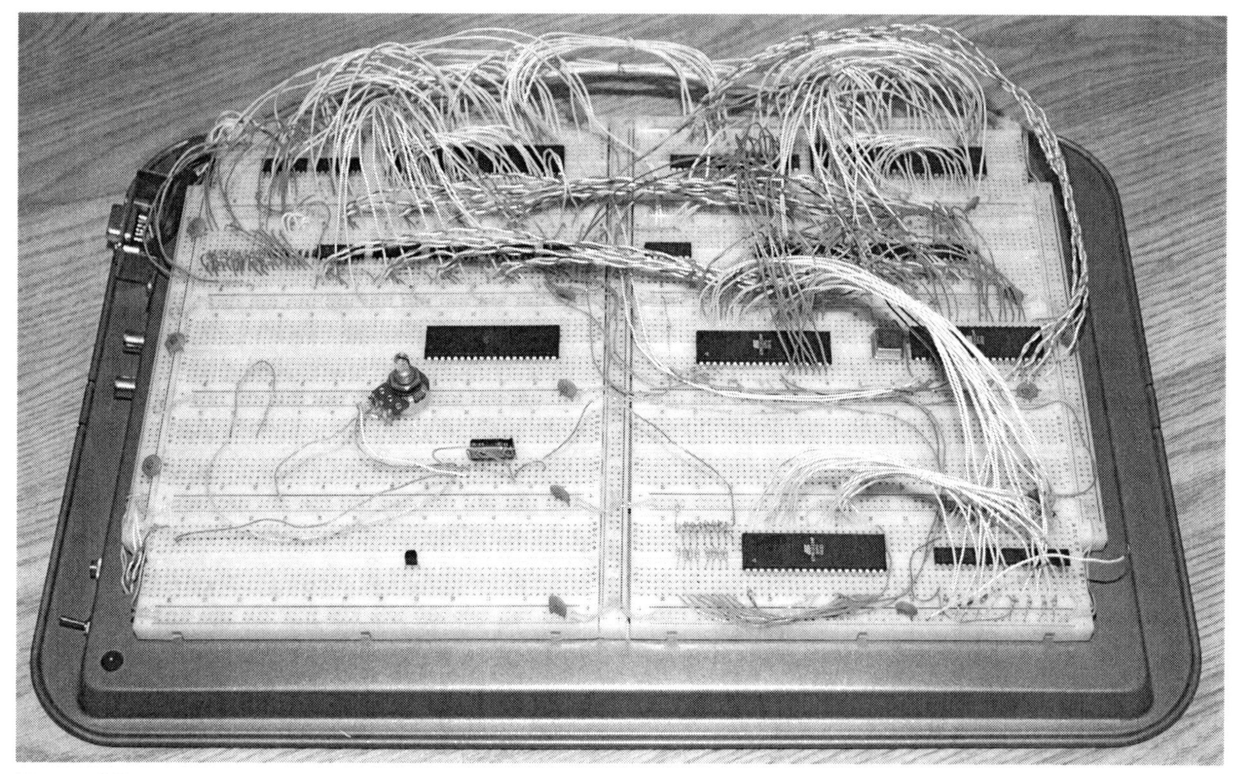

Figure I-7 *Breadboard circuits can get large.*

boards. Dead VCRs, fried TVs, baked radios, and even that broken coffee maker will have a pile of usable components on the circuit board. A small radio PCB might have 200 semiconductors soldered to it, and at 50 cents apiece, that adds up fast. You may never use all the components, but on a dreary day when you are in your "mad scientist" laboratory in need of some oddball resistor value, a box of scrap circuit boards is great to have.

I keep several large boxes full of circuit boards that I find, and I often find that they yield most of the common parts I need and often have hard-to-find or discontinued ICs that I need when working on older schematics. Figure I-8 shows one of the 20 or more large boxes of scrap PCBs I have collected over the years. Removing the parts from an old circuit board is easy, especially the simple two- or three-pin parts such as capacitors and transistors. For the larger ICs with many pins, you will need a desoldering tool, also known as a "solder sucker." A low-budget soldering iron (34 W or higher) with a blunt tip and a solder sucker can make easy work of pulling parts from old PCBs. Figure I-9 shows a hand-operated solder sucker removing an eight-pin IC from an old VCR main board. To operate a solder sucker, you press down on the loading lever, heat up the pad to desolder, and then press a button to suck up the solder away from that pad.

I prefer to desolder using a cheap iron with a higher wattage and fatter tip than used for normal work because it heats up the solder faster, reaching both sides of the board easier. There are other tools that can be used to desolder parts, such as desoldering wicks, spade tips, and even pump vacuums, but the $10 solder sucker has always done the job for me, even on ICs with as many as 40 pins, without any problems at all. When you really start collecting parts, you will find that a giant bowl of resistors is more of a pain that having to desolder a new one, so some type of organization will be necessary. You will soon discover that there are many common parts when it comes to resistors, capacitors, transistors,

Figure 1-8 *Old PCBs are a gold mine of parts.*

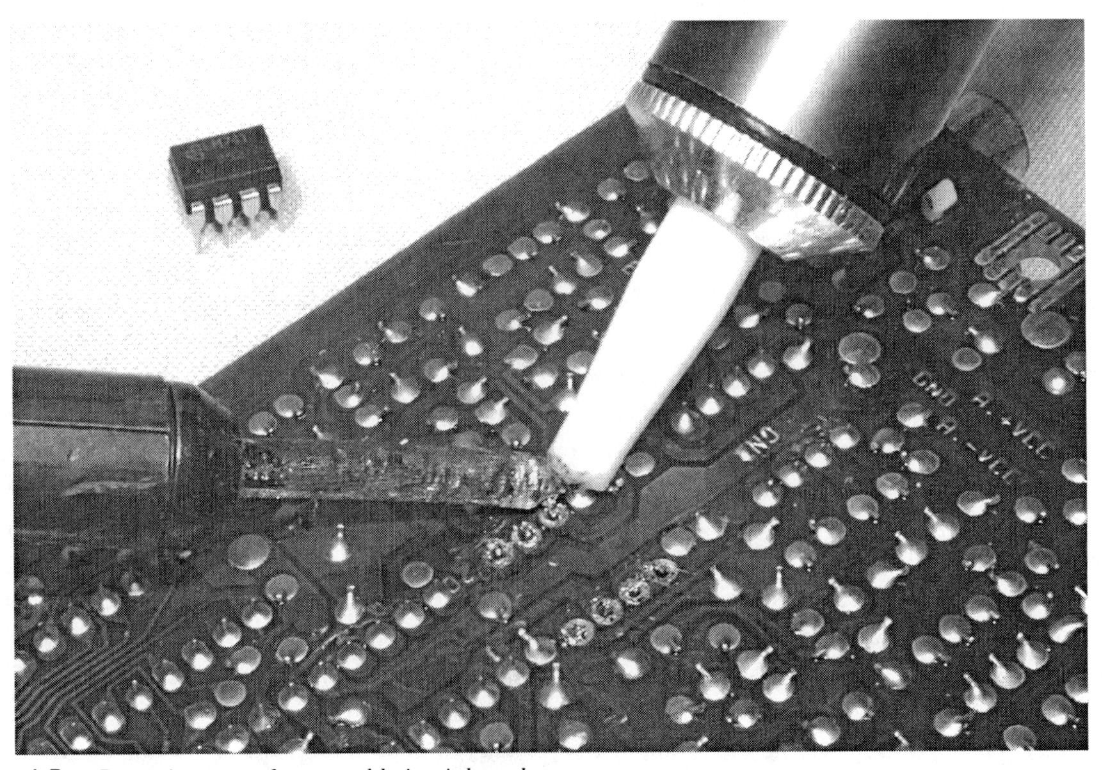

Figure 1-9 *Removing parts from an old circuit board.*

Figure I-10 *Organization makes finding parts a snap.*

and ICs, so having them sorted will make it very easy to find what you need. Storage bins such as the ones shown in Figure I-10 are perfect for your electronic components, and you can easily fill 100 small drawers with various parts, so purchase a few of them to get started.

I have an entire closet full of these component drawers, and I store the larger parts or PCBs in plastic tubs. It's a rare day when I can't find the parts I need, even for a retro project that needs some long-discontinued component. Of course, there are always times when you will want new parts or need something special, so one of the many online sellers will be glad to take your money and send you the part in a few days.

The Resistor

If you are new to this hobby, then you probably have seen a few schematics and thought that they made about as much sense as cave hieroglyphics. Don't worry, though; knowledge of schematics will come as you use them more and start to decode the electronic-component datasheets. If you want to take the fast track, then you might consider getting a book on basic electronics to get you kick-started, but for those who want to learn as you go, here are a few of the basics you will need to complete the projects in this book.

Resistors like the ones shown in Figure I-11 are the most basic of the semiconductors you will be using, and they do exactly what their name implies—they resist the flow of current by exchanging some current for heat that is dissipated through the body of the device. On a large circuit board, you could find hundreds of resistors populating the board, and even on tiny circuit boards with many surface-mounted components, resistors usually will make up the bulk of the semiconductors. The size of the resistor generally determines how much heat it can dissipate, and the resistor will be rated in watts, with ¼ and ⅛ W being the most common type you will work with (the two bottom resistors shown in the figure). Resistors can become very large and will require ceramic-based bodies, especially if they are rated for several watts or more, such as the 10-W unit

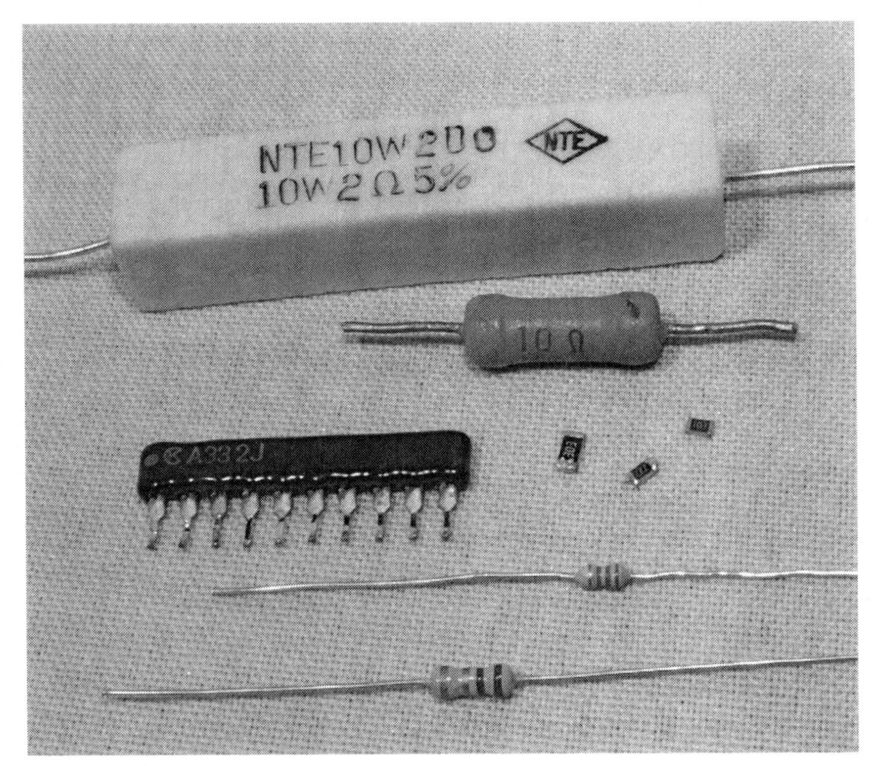

Figure I-11 *Resistors with fixed values.*

shown at the top of Figure I-11. To save space, some resistors come in packs, such as the one in the figure that has multiple legs.

Because of the recent drive to make electronics more "green" and power conservative, large, power-wasting resistors are not common in consumer electronics these days because it is more efficient to convert amperage and voltage using some type of switching power supply or regulator rather than by letting a fat resistor burn away the energy as heat. On the other hand, small-value resistors are very common, and you will find yourself dealing with them all the time for simple tasks such as driving a light-emitting diode (LED) with limited current, pulling up an input pin to a logical 1 state, biasing a simple transistor amplifier, and thousands of other common functions. On most common axial-lead resistors, such as the ones you will use most often in your projects, the value of the resistor is coded onto the device in the form of four colored bands that tell you the resistance in ohms. Ohmage is represented using the Greek capital letter omega (Ω) and often will be omitted for values over 99 Ω, which will be stated as 1K, 15K, 47K, or some other number followed by the letter K, indicating the value is in kiloohms (thousands of ohms). Similarly, for values over 999K, the letter *M* will be used to show that 1M is actually 1 megaohm, or 1 million Ω. In a schematic diagram, a resistor is represented by a zigzag line segment, as shown in Figure I-12, and will either have a letter and a number such as R1 or V3 relating to a parts list or simply will have the value printed next to it, such as 1M, or 220 Ω. The schematic symbol on the left of Figure I-12

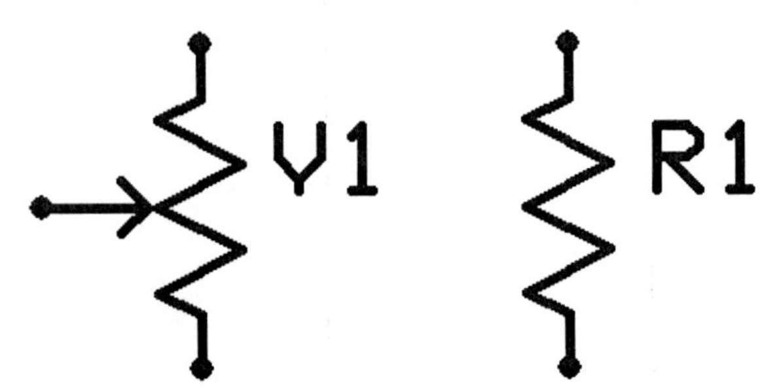

Figure I-12 *Variable (left) and fixed (right) resistor symbols.*

represents a variable resistor, which can be set from 0 Ω to the full value printed on the body of the variable resistor.

A variable resistor is also known as a *potentiometer*, or *pot*, and it can take the form of a small circuit board–mounted cylinder with a slot for a screwdriver or a cabinet-mounted can with a shaft exiting the can for mating with some type of knob or dial. When you crank up the volume on an amplifier with a knob, you are turning a potentiometer. Variable resistors are great for testing a new design because you can just turn the dial until the circuit performs as you want it to and then remove the variable resistor to measure the impedance (resistance) across the leads to determine the best value of fixed resistor to install. On a variable resistor, there are usually three leads—the outer two connect to the fixed carbon resistor inside the can, which gives the variable resistor its value, and a center one connects to a wiper, allowing the selection of resistance from zero to full. Several common variable resistors are shown in Figure I-13, with the top-left unit dissected to show the resistor band and wiper.

As mentioned earlier, most fixed-value resistors will have four color bands painted around their bodies that can be decoded into a value, as shown in Table I-1. At first, this may seem a bit illogical, but once you get the hang of color-band decoding, you will be able to recognize most common values at first glance without having to refer to the chart.

There almost always will be either a silver or gold band included on each resistor, and this will indicate the end of the color sequence and will not become part of the value. A gold band indicates the resistor has a 5 percent tolerance (margin of error) in the value, so a 10K resistor could end up being anywhere from 9.5K to 10.5K in value, although in most cases it will be very accurate. A silver band indicates that the

Figure I-13 *Common variable resistors.*

TABLE I-1 Resistor Color-Code Chart

Color	*1st Band*	*2nd Band*	*3rd Band*	*Multiplier*
Black	0	0	0	1Ω
Brown	1	1	1	10Ω
Red	2	2	2	100Ω
Orange	3	3	3	1KΩ
Yellow	4	4	4	10KΩ
Green	5	5	5	100KΩ
Blue	6	6	6	1MΩ
Violet	7	7	7	10MΩ
Grey	8	8	8	
White	9	9	9	0.1
Gold				0.01
Silver				

tolerance is only 10 percent, but I have yet to see a resistor with a silver band that was not on a circuit board that included vacuum tubes, so forget that there is even such a band! Once you ignore the gold band, you are left with three color bands that can be used to determine the exact value as given in Table I-1.

So let's assume that we have a resistor with the color bands brown, black, red, and gold. We know that the gold band is the tolerance band, and the first three will indicate the values to reference in the chart. Doing so, we get 1 (brown), 0 (black), and 100 Ω (red). The third band is the multiplier, which would indicate that the number of zeroes following the first to values will be 2, or the value is simply multiplied by 100 Ω. This translates to a value of 1000 Ω, or 1K (10 × 100 Ω). A 370K resistor would have the colors orange, violet, and yellow followed by a gold band. You can check the value of the resistor when it is not connected to a circuit simply by placing your multimeter on the appropriate resistance scale and reading back the value. I do not want to get too deep into electronics formulas and theory here because there are many good books dedicated to the subject, so I will simply leave you with two basic rules regarding the use of resistors: (1) Put them in series to add their values together, and (2) put them in parallel to divide them. This second simple rule works great if you are in desperate need of a 20K resistor, for instance, but can find only two 10K resistors to put in series. In parallel, they will divide down to 5K. Now you can identify the most common semiconductor that is used in electronics today, the resistor, so we will move ahead to the next most common semiconductor, the capacitor.

The Capacitor

A *capacitor* in its most basic form is a small rechargeable battery with a very short charge and discharge cycle. Where a typical AAA battery may be able to power an LED for a full year, a capacitor of similar size will power it for only a few seconds before its energy is fully discharged. Because capacitors can store energy for a predictable duration, they can perform all kinds of useful functions in a circuit, such as filtering ac waves, creating accurate delays, removing impurities from a noise signal, and creating clock and audio oscillators. Because a capacitor is basically a battery, many of the large ones look much like batteries with two leads connected to

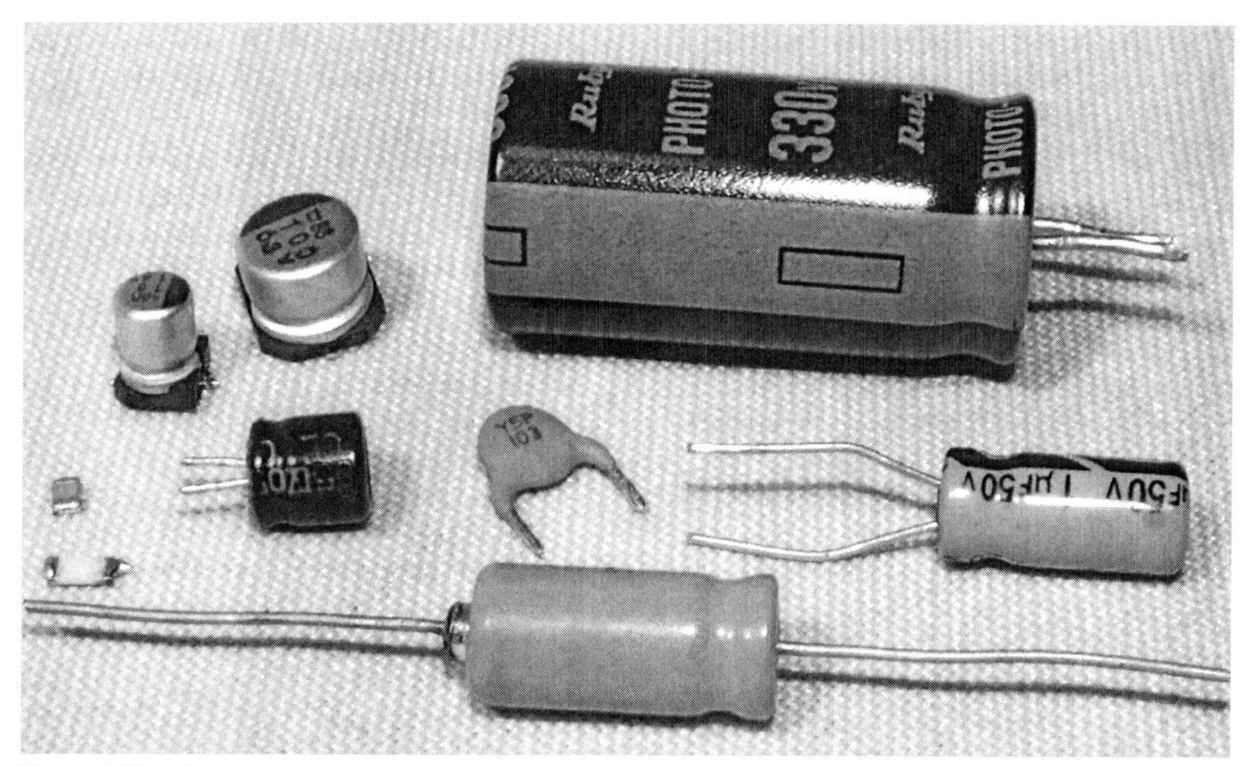

Figure I-14 *Various common capacitors.*

one side of a metal can. As shown in Figure I-14, there are many sizes and shapes of capacitors, some of which look like small batteries.

Just like resistors, capacitors can be as large as a garbage can or as small as a grain of rice—it really depends on the value. The larger devices can store a lot more energy. Unlike batteries, some capacitors are nonpolarized, and they can be inserted into a circuit regardless of current flow, whereas some cannot. The two different types of capacitors are shown by their schematic symbols in Figure I-15, C1 being a nonpolarized type and C2 a polarized type. Although there are always exceptions to the rules, generally, the disk-style capacitors are nonpolarized, and the larger can-style electrolytic types are polarized. An obvious indicator of a polarized capacitor is the negative markings on the can, which can be seen clearly in the larger capacitor shown at the top right of Figure I-14.

Another thing that capacitors have in common with batteries is that polarity is very important when inserting polarized capacitors into a circuit. If you install an electrolytic capacitor in reverse and attempt to charge it, the part likely will heat up and release the oil contained inside the case, causing a circuit malfunction or dead short. In the past, electrolytic capacitors did not have a pressure-release system and would explode like firecrackers when overcharged or installed in reverse, leaving behind a huge mess of oily paper and a smell that was tough to forget. On many capacitors, especially the larger can-style ones, the voltage rating and capacitance value are

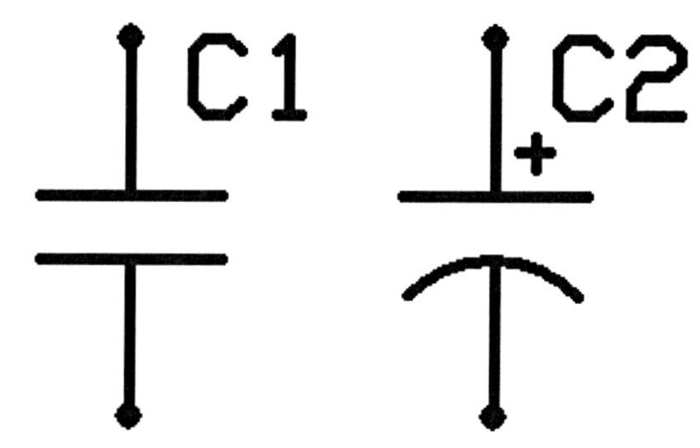

Figure I-15 *Capacitor symbols.*

simply stamped on the case. A capacitor is rated in voltage and in farads, which defines the capacitance of a dielectric for which a potential difference of 1 V results in a static charge of 1 C.

This may not make a lot of sense until you start messing around with electronics, but you will soon understand that, typically, the larger the capacitor, the larger the farad rating will be, and thus the more energy it can store. Since a farad is quite a large value, most capacitors are rated in microfarads (μF), such as the typical value of 4700 μF for a large electrolytic filter capacitor and 0.1 μF for a small ceramic-disk capacitor. Picofarads (pF) are also used to indicate very small values, such as those found in many ceramic capacitors or adjustable capacitors used in rf circuits (a picofarad is one-millionth of a microfarad).

On most can-style electrolytic capacitors, the value is simply written on the case and will be stated in microfarads and voltage, along with a clear indication of which lead is negative. Voltage and polarity are very important in electrolytic capacitors, and they always should be inserted correctly, with a voltage rating higher than necessary for your circuit. Ceramic capacitors usually will have only the value stamped on them if they are in picofarads for some reason, and often no symbol will follow the number, just the value. Normally, ceramic capacitors will have a

TABLE I-2 Ceramic Capacitor Value Chart

Value	*Marking*	*Value*	*Marking*	*Value*	*Marking*
10 pf	10 or 100	0.001 uF	102	0.10 uF	104
12 pf	12 or 120	0.0012uF (1200pf)	122	0.12 uF	124
15 pf	15 or 150	0.0015uF	152	0.15 uf	154
18 pf	18 or 180	0.0018 uF (1800pf)	182	0.18 uF	184
22 pf	22 or 220	0.0022uF	222	0.22 uF	224
27 pf	27 or 270	0.0027uF	272	0.27 uF	274
33 pf	33 or 330	0.0033 uF	332	0.33 uF	334
39 pf	39 or 390	0.0039uF	392	0.39 uF	394
47 pf	47 or 470	0.0047uF	472	0.47 uF	474
58 pf	58 or 580	0.0056uF	562	0.56 uF	564
68 pf	68 or 680	0.0068uF	682	0.68 uF	684
82 pf	82 or 820	0.0082uF	822	0.82 uF	824
100 pf	101	0.01 uF	103	1uF	105 or 1uf
120 pf	121	0.012 uF	123		
150 pf	151	0.015 uF	153		
180 pf	181	0.018 uF	183		
220 pf	221	0.022 uF	223		
270 pf	271	0.027 uF	273		
330 pf	331	0.033 uF	333		
390 pf	391	0.039 uF	393		
470 pf	471	0.047 uF	473		
560 pf	561	0.056 uF	563		
680 pf	681	0.068 uF	683		
820 pf	821	0.082 uF	823		

three-digit number that needs to be decoded into the actual value, and this confusing scheme works as shown in Table I-2.

Who knows why they just don't write the value on the capacitor? I mean, it would have the same number of digits as the code! Oh well, you get used to seeing these codes, just like resistor color bands, and in no time you will easily recognize the common values, such as 104, which would indicate a 0.1-μF value according to the chart. Capacitors behave just like batteries when it comes to parallel and series connections, so in parallel, two identical capacitors will handle the same voltage as a single unit but double their capacitance rating, and in series, they will have the same capacitance rating as a single unit but can handle twice the voltage. Thus, if you need to filter a really noisy power supply, you might want to install a pair of 4700-μF capacitors in parallel to end up with a capacitance of 9400 μF. When installing parallel capacitors, make sure that the voltage ratings of all the capacitors used are higher than the voltage of that circuit, or there will be a failure—an ugly, noisy, smelly failure!

The Diode

Diodes allow current to flow through them in one direction only, so they can be used to rectify ac into direct current (dc), block unwanted current from entering a device, protect a circuit from a power reversal, and even give off light in the case of LEDs. Figure I-16 shows various sizes and types of diodes, including an easy recognizable LED and the large full-wave-rectifier module at the top. A full-wave rectifier is just a block containing four large diodes inside.

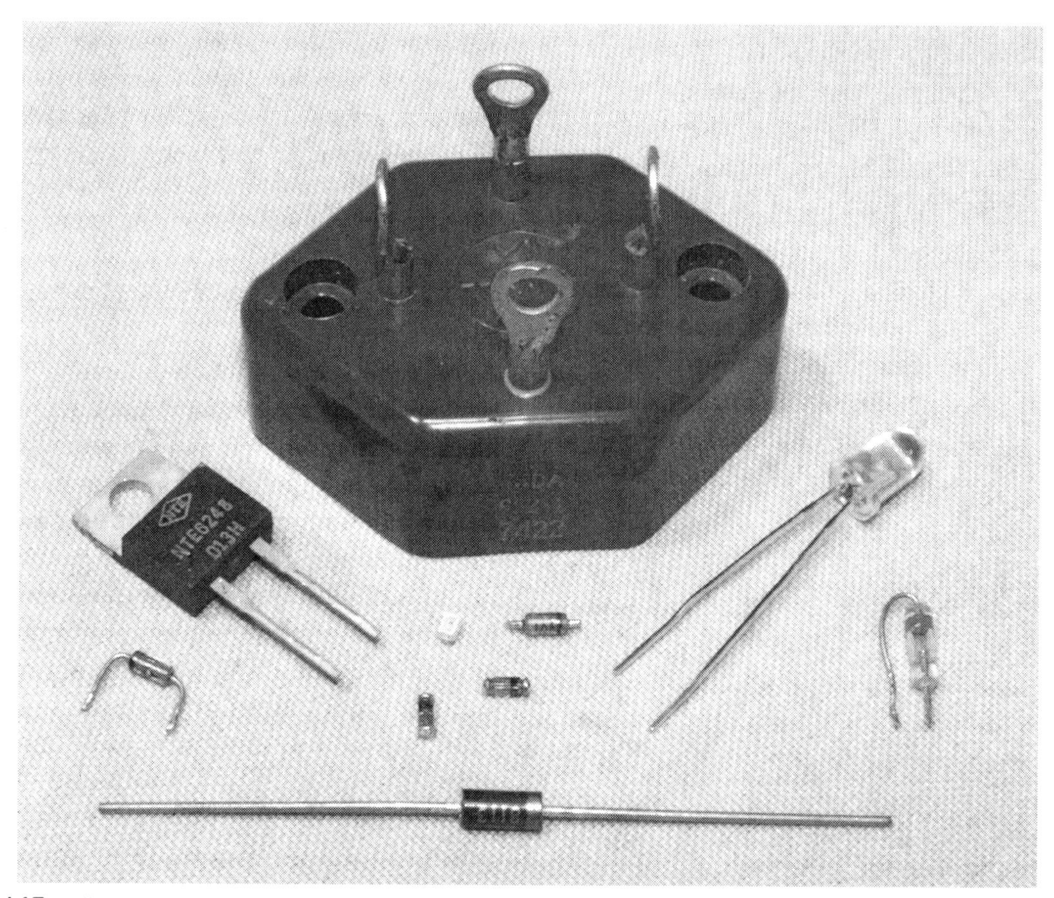

Figure I-16 *Several styles of diodes, including an LED.*

As with most other semiconductors, the size of the diode is usually a good indication of how much current it can handle before failure, and this information will be specified by the manufacturer by referencing whatever code is printed on the diode to some datasheet. Unlike resistors and capacitors, there is no common mode of identifying a diode unless you get to know some of the most common manufacturers' codes by memory, so you will be forced to look up the datasheet on the Internet or in a cross-reference catalog to determine the exact value and purpose of unknown diodes.

For example, the NTE6248 diode shown in Figure I-16 in the TO220 case (left side of figure) has a datasheet that indicates that it is a Schottky barrier rectifier with a peak reverse-voltage maximum of 600 V and a maximum forward current rating of 16 A. Datasheets will tell you everything you need to know about a particular device, and you should never exceed any of the recommended values if you want a reliable circuit. The schematic symbol for a diode is shown in Figure I-17, D1 being a standard diode and the other an LED (the two arrows represent light leaving the device).

The diode symbol shows an arrow (anode) pointing at a line (cathode), and this will indicate which way current flows (from the anode to the cathode, or in the direction of the arrow). On many small diodes, there will be a stripe painted around the case to indicate which end is the cathode, and on LEDs, there will be a flat side on the case nearest the cathode lead. LEDs come in many different sizes, shapes, and wavelengths (colors) and have ratings that must not be exceeded to avoid damaging the device. Reverse voltage and peak forward current are very important values that must not be exceeded when powering LEDs, or damage will easily occur. At the same time, though, you will want to get as close as possible to the maximum value if your circuit demands full performance from an LED, so read the datasheets on the device carefully. Larger diodes used to rectify ac or control large current may need to be mounted to the proper heat sink to operate at their rated values, and often the case style will be a clear indication owing to the metal backing or mounting hardware that may come with the device. Unless you know how much heat a certain device can dissipate in open air, your best bet is to mount it to a heat sink if it was designed to be installed that way. As with most semiconductors, there are thousands of types and sizes of diodes, so make sure that you are using a part rated for your circuit, and double-check the polarity of the device before you turn on the power for the first time.

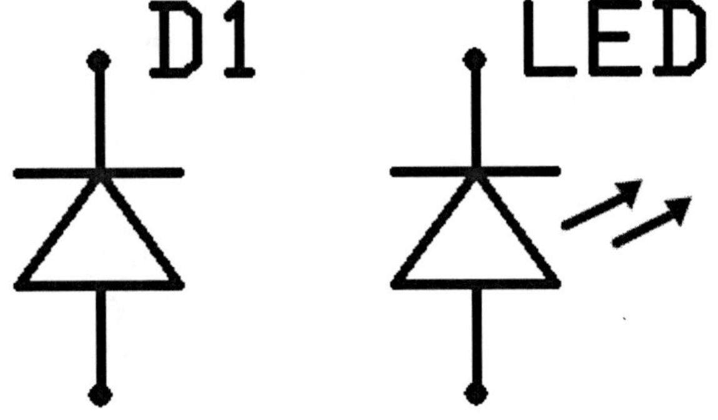

Figure I-17 *Diode schematic symbol (LED on the right).*

The Transistor

A *transistor* is one of the most useful semiconductors available and often the building block for many larger ICs and components such as logic gates, memory, and microprocessors. Before transistors became widely used in electronics, simple devices such as radios and amplifiers would need huge wooden cabinets, consume vast amounts of power, and emit large, wasteful quantities of heat owing to the use of vacuum tubes. A vacuum tube–based computer called ENIAC was built that used 17,468 vacuum tubes, 7200 crystal diodes, 1500 relays, 70,000 resistors, and 10,000 capacitors and had more than 5 million hand-soldered joints. It weighed 30 tons and was roughly 8 × 3 × 100 ft in size and consumed 150 kW of power! A simple computer that would rival the power of this power-hungry monster could easily be built on a few square inches of perforated board using a few dollars in parts today by any electronics hobbyist, thanks to the transistor. A transistor is really just a switch that can control a large amount of current by switching a small amount of current, thus creating an amplifier. Several common types and sizes of transistors are shown in Figure I-18.

Depending on how much current a transistor is designed to switch, it may be as small as a grain of rice or as large as a hockey puck and require a massive steel heat sink or fan to operate correctly. There are thousands of transistor types and sizes, but one thing most of them have in common is that they will have three connections that can be called the *collector*, the *emitter*, and the *base* and will be represented by one of the two schematic symbols shown in Figure I-19.

The emitter (E), base (B), and collector (C) on both the NPN and PNP transistors do the same job. The collector/emitter current is controlled by the current flowing between the base and emitter terminals, but the flow of current is opposite in each device. Today, most transistors are NPN owing to the fact that it is easier to manufacture a better NPN transistor than a PNP, but there are still times when a circuit may use a PNP transistor owing to the direction of current or in tandem with an NPN transistor to create a matched pair. There is enough transistor theory to cover 10 books of this size, so I will condense that information to help you understand the very basics of transistor operation. As a simple switch, a transistor can be thought of as a relay with no mechanical parts. You can turn on a high-current load such as a light or motor with a very weak current such as the output from a logic gate or

Figure I-18 *Various common transistors.*

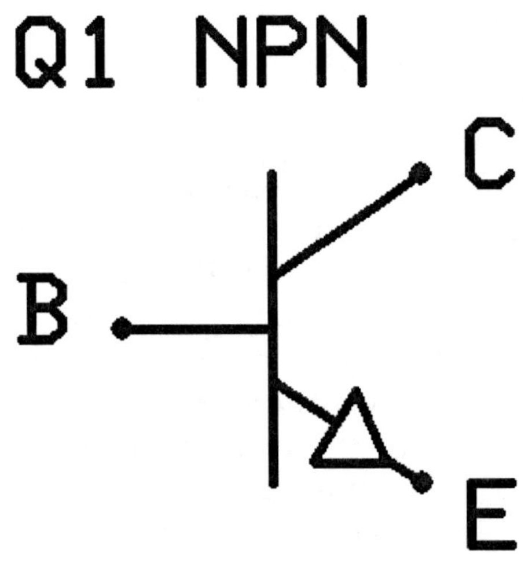

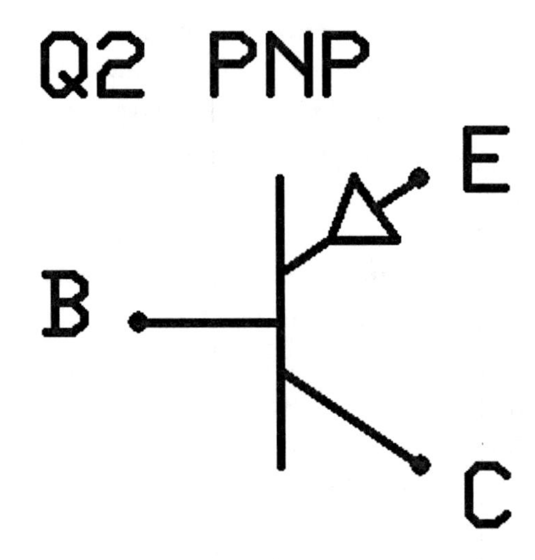

Figure I-19 *NPN and PNP transistor schematic symbols.*

light-sensitive photocell. Switching a large load with a small load is very important in electronics, and transistors do this perfectly and at speed that a mechanical switch such as a relay could never come close to achieving. A audio amplifier is nothing more than a very fast switch that takes a very small current such as the output from a CD player and uses it as the input into a fast switch that controls a large current such as the dc power source feeding the speakers. Almost any transistor can easily operate well beyond the frequency of an audio signal, so they are perfectly suited for this job. At much higher frequencies, such as those used in radio transmitters, transistors do the same job of amplification but are rated for much higher frequencies, sometimes into the gigahertz range.

Another main difference between the way a mechanical switch and a transistor works is the fact that a transistor is not simply an on or off switch. It can operate as an *analog* switch, varying the amount of current switched by varying the amount of current entering the base of the transistor. A relay can turn on a 100-W light bulb if a 5-V current is applied to the coil, but a transistor could vary the intensity of the same light bulb from zero to full brightness depending on the voltage seen at the base. As with all semiconductors, the transistor must be rated for the job you intend it to do, so maximum current, switching voltage, and speed are things that need to be considered when choosing the correct part. The datasheet for a very common NPN transistor, the 2N2222 (which can be substituted for the 2N3904 often used in this book), is shown in Figure I-20.

From this datatsheet, you can see that this transistor can switch about half a watt (624 mW) with a voltage of 6 V across the base and emitter junction. Of course, these are maximum ratings, so you might decide that the transistor will work safely in a circuit if it had to switch on a 120-mW LED from a 5-V logic-level input at the base. As a general rule, I would look at the maximum switching current of a transistor and never ask it to handle more than half the rated maximum value, especially if it is the type of transistor designed to be mounted to a heat sink. The same thing applies to maximum switching speed—don't expect a 100-MHz transistor to oscillate at 440 MHz in an rf transmitter circuit because it will have a difficult enough time just reaching the 100-MHz level.

Amplifier Transistors
NPN Silicon

P2N2222A

MAXIMUM RATINGS

Rating	Symbol	Value	Unit
Collector–Emitter Voltage	V_{CEO}	40	Vdc
Collector–Base Voltage	V_{CBO}	75	Vdc
Emitter–Base Voltage	V_{EBO}	6.0	Vdc
Collector Current — Continuous	I_C	600	mAdc
Total Device Dissipation @ T_A = 25°C Derate above 25°C	P_D	625 5.0	mW mW/°C
Total Device Dissipation @ T_C = 25°C Derate above 25°C	P_D	1.5 12	Watts mW/°C
Operating and Storage Junction Temperature Range	T_J, T_{stg}	–55 to +150	°C

THERMAL CHARACTERISTICS

Characteristic	Symbol	Max	Unit
Thermal Resistance, Junction to Ambient	$R_{\theta JA}$	200	°C/W
Thermal Resistance, Junction to Case	$R_{\theta JC}$	83.3	°C/W

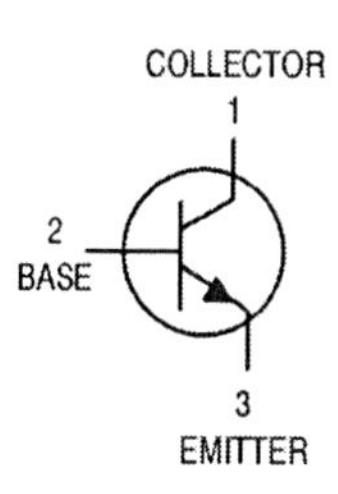

Figure I-20 *Datasheet for the common 2N2222 NPN transistor.*

RTFM—Read the Flippin' Manual!

Let's face it, "evil geniuses" don't often read manuals and frequently prefer to learn by trial and error, which really is the best way to learn most of the time. When it comes to determining the specs on an electronic component with more than two pins, you really have no choice but to read the datasheet to figure out how it works and to avoid letting out the smoke. This is especially true for transistors and ICs because you really have no idea what is inside the "black box" without the datasheet. That eight-pin IC may be just a simple timer, or it could be a state-of-the-art 100-MHz microprocessor with a built-in USB port and a video output. Without the datasheet, you would never know. There are also times when you may have found some cool schematic on the Internet, and it has a parts list that you determine contains mostly discontinued transistors, so you will have to compare the datasheets to find suitable replacements. Once you learn the basics, you will be able to make smart substitutions for almost any part.

You can dig up just about any datasheet on any device, even those that have been off the market for decades, so learn how to find them, and you will never be in the dark when it comes to component specs. The datasheet for the very common 2N2222 transistor was found by typing, you guessed it, "2N2222 datasheet" into Google (Figure I-21) and looking for the PDF file. Although there are a few bogus datasheet servers out there that try to suck you into joining so that they can spam your e-mail address for life, most of the time you will find a datasheet with little or no fuss. If you know the manufacturer of the

Figure I-21 *The Internet is your friend.*

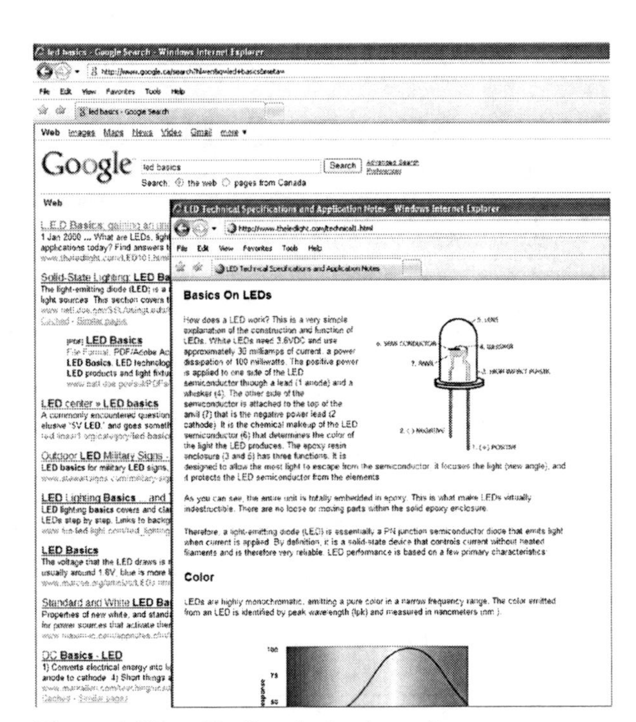

Figure I-22 *Finding the basics online.*

device, try its Web site first, or try a large online electronics supply site such as Digikey.com because it will have online datasheets for thousands of components. The Internet is also a great source for general information on electronics, so if you find yourself needing to know the basics, consult your favorite search engine.

The wealth of information found by searching "LED basics" was almost endless. Even the first site (Figure I-22) had more than enough information on the LEDs presented in a very easy-to-understand format. The fact is that almost all the information you will need when learning electronics can be found on the Internet with a little patience.

Asking for Help

When the Internet fails to provide you with answers or you really feel that you need guidance from those who may know the answers, there are countless forums that you can join and look for help or discuss your projects with other "evil geniuses." As with all things in life, forums have their own special rules of etiquette, so before you jump in and scream "Help me!", please take the time to read the posting rules and consider the following: Most of the knowledgeable people on a forum who would consider answering your question(s) are doing so on their own time, just to be nice. They do this because they remember what it felt like to be starting out and in need of a little guidance to make a project a success. If you have not bothered to cover the basics or given much effort to solving your own problems first, you will get zero respect in a forum and either be ignored or "spanked" publicly.

Here is how *not* to ask for help in a forum (I see this a lot):

> Urgent: can't program the microcontroller. Pls help me, urgent! i am a noob and don't know how 2 get the code in2 the avr. Whut do I need 2 buy? Can some1 pls send me the file?

This type of post may never get a response for many reasons. First of all, why would your problem be of concern to anyone else? Most people on a forum also have a life and a job, so their first thought will be, "If your problem is so urgent, pay someone to solve it." Be patient, and remember that if someone is kind enough to help you in the next few days, then you are very lucky. Also, making multiple posts of the same or similar request in other threads hoping that more posts from you will encourage responses likely will get you into trouble with moderators and regular members. It may take days for a response, but remember that everyone else has urgent things in their lives to look after before taking the time to help you and others.

The next reason why most forum members will ignore this kind of post is because of the sloppy use of language. Posting like this would be the same as showing up for a formal job interview wearing your garage clothes and is a clear indication of laziness. The message also indicates that the poster has done nothing to help himself or herself and expects to be "spoonfed" by those who have done the groundwork or have invested in many years of formal and informal education. The question also has little information, so the forum members are left to guess.

Here is a much better way to ask the same question and likely get a response:

> Project X—Question regarding the AVR selection. Hello everyone. I'm working on the Mind Reader project and was wondering if the AVR644p could be used to replace the AVR324p? I have read the datasheets on both chips but just want to make sure that there aren't some issues with this chip that I should know about. Has anyone out there compiled a HEX file for this device yet? If not, I'm probably going to give it a try and upload the file if it works. Thanks for your help.

Now, the poster has asked a clear question that indicates both the project and the part numbers, as well as what he or she has done to help himself or herself first. People who are willing to try to solve their own problems first and then ask politely for additional help are welcomed in a forum because they are likely the type of person to come back later and help others when they make the transition from newcomer to experienced hobbyist. Consider these points before asking for help online.

Tools of the Trade

Once you have a good supply of junk to mess around with and a basic understanding of the parts and schematic symbols, you can begin to turn your ideas into reality. Of course, you will need a few basic tools, which can be purchased at most electronics suppliers. Much like the breadboard, a soldering iron is the workhorse in the electronics industry. You can create and test your circuit on a breadboard, but to move it to a more permanent home, you will have to solder those components down. I like to have two soldering irons for this hobby—one is a cheap unit to use when unsoldering parts from scrap boards, and the other is a better, heat-controlled unit with interchangeable tips for fine work.

The ability to control the heat and replace the tips on a soldering iron such as the one shown in Figure I-23 makes it easy to work on all types of projects. For delicate surface-mounted components, you need a sharp point and low heat, but for soldering down a large heat sink, you need a blunt tip and lots of heat, so the ability to turn a dial and switch the tip is very handy. Also shown in the figure is the holster and sponge bath for cleaning the tip. You certainly can get away with using a $10 soldering iron for most of the projects in this book, but if you plan to dive deep into this hobby, then consider investing in a quality soldering station.

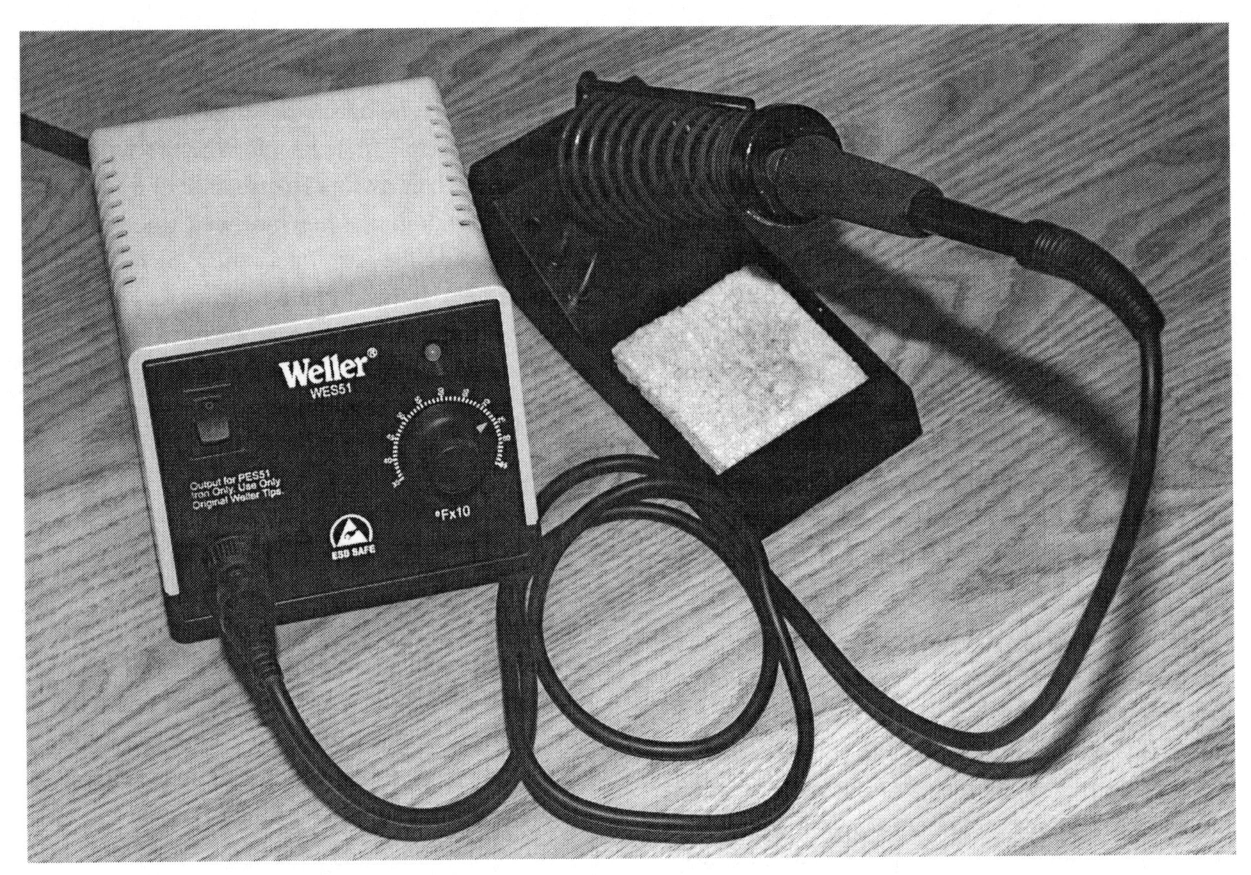

Figure I-23 *A midrange heat-controlled soldering iron.*

Unless everything you build is going to run from a 9-V battery, you will need some type of adjustable power supply like the one shown in Figure I-24. Often, a circuit will need two different voltages, so a dual supply makes it easy to just dial up the voltage you need. These test-bench power supplies also allow you to set the current, so you can help to eliminate fried components by slowly adjusting the voltage while you watch for spikes in the current. If you accidentally reverse the power pins on your last microcontroller and drop on a 1-A "wall wart," you most certainly will see some fireworks. If you limit the current on an adjustable supply and slowly turn up the voltage, chances are very high that you will catch the error before any damage is done to the chip. Having an amp meter also lets you figure out how much power your circuit needs when it comes time to choose a regulator. An adjustable power supply is another one of those "must have" devices for the electronics hobby.

A multimeter such as the one shown in Figure I-25 also will be needed to measure component values, check voltages, and test your circuits. A very basic multimeter will measure voltage, current, and resistance, and a more advanced model also might include a frequency counter, capacitance tester, transistor checker, and even some basic graphing functions. You will use your multimeter to check voltage and measure resistance most of the time, so even a basic model will be fine for most work.

If you really want to run with this hobby or learn electronics on a professional level, you eventually will want an oscilloscope (Figure I-26) for your bench. A modern oscilloscope can measure just about anything and then display it on a high-resolution color screen or download the

Figure 1-24 *A variable dual power supply.*

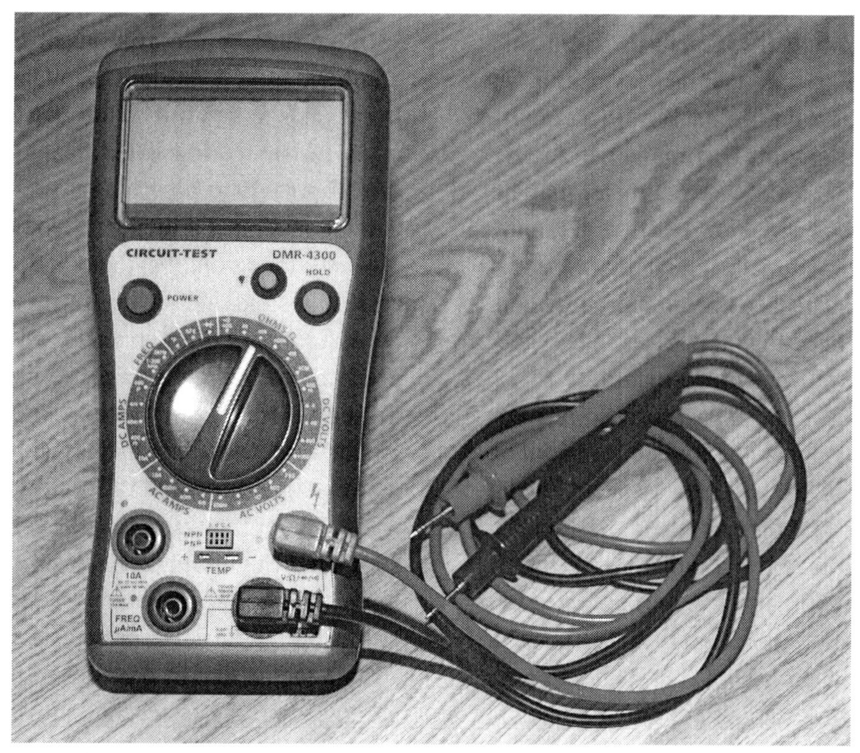

Figure 1-25 *A basic digital multimeter.*

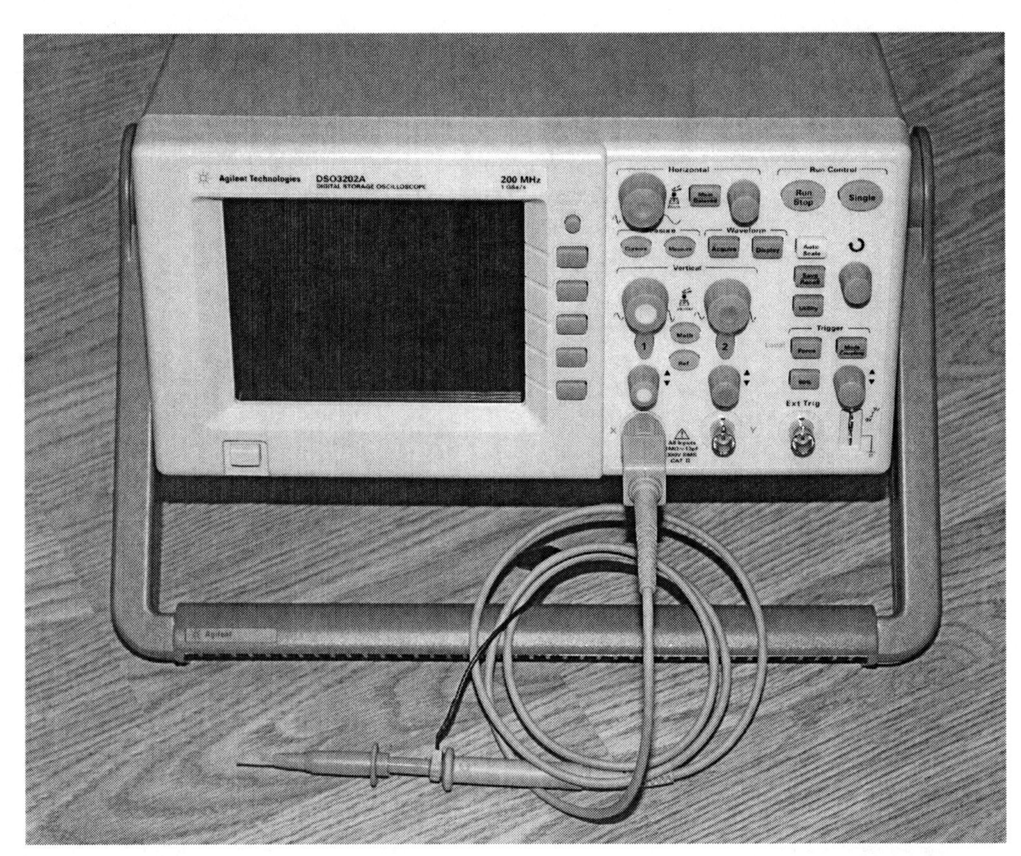

Figure 1-26 *An oscilloscope for more advanced work.*

results to your computer for analysis. Even an older tube-based "scope" can be very useful when you need to see instant feedback from a circuit on a screen. As your circuits become larger, faster, and more complex, a simple multimeter may not be enough to debug problems because it cannot respond in real time or detect such high-speed voltage changes. This becomes important when designing high-speed digital devices that expect a certain sequence of digital information to be transmitter or received. An oscilloscope lets you "dig deep" into a signal and look at it one bit at a time until you find the bug. You also can display analog signals in real time, compare multiple signals, and even save the results to a PC for further analysis. Although you will not need an oscilloscope to build the projects in this book, I could not have done many of them without one, so when you start designing your own circuits, a scope is invaluable.

In the time before millions of transistors were crammed onto a single IC, you had a few basic logic gates, a handful of transistors, and your usual resistors and capacitors. Now the selection of components is so vast that printing a supplier's catalog is almost a dream. If you can dream up a use for 1s and 0s, there are probably 10 companies making a chip for it, so the selection is massive, and often, because of mass-scale production, finding a single IC can be very difficult. Maybe you just tore into a VCR and isolated the on-screen display chip, found the datasheet, and thought, "Cool, I want that for my robot." After a bit of a search, you find that the supplier's minimum order quantity is 100,000 units, and the programmer needed to make the chip work is the same price as a small car. The simple fact is that as electronics become more complex and the faster the industry moves, the less chance us hobbyists have of sourcing the latest parts. Sure, transistors, resistors, and the basic components will always be available, but complex ICs designed for consumer devices either will not be available in small quantities or will have packages so small that you need a microscope just to see all the pins. Enter the microcontroller.

A *microcontroller* is like a blank IC ready to be made into whatever you want it to be. A $5 microcontroller can take on the function of a 16-input gate or become something as complex as a fully functional Pong game that will display video on your television. A microcontroller is a chip that contains a processor, some memory, and a few peripherals such as serial ports or even a high-speed USB port. There are even microcontrollers that have MP3 decoders and high-resolution video generators onboard. Although learning to program a microcontroller is almost a hobby in itself, it does open up a door that allows you to create just about any function on a single IC so that you don't have to worry about what the market is doing or if a chip will be available. Sure, there are limits, but not that many, and microcontrollers are king in the hobby world of electronics. Figure I-27 shows a few microcontrollers with varying pin counts and a programmer from Microchip as well as Parallax.

To make a microcontroller do your bidding, you first write code on your PC using C, Assembly, or Basic, and then you compile and send your program to the programmer, which fits it into the microcontroller onboard Flash memory. If your code works properly, your microcontroller now becomes your own custom IC. Microcontrollers typically cost between $5 and $20 depending on Flash memory size and onboard peripherals, and programmers cost between $50 and $200 depending on chip support and debug functionality. There are even Web sites that show you how to build your own programmers for a few bucks, so microcontrollers are certainly well within anyone's budget.

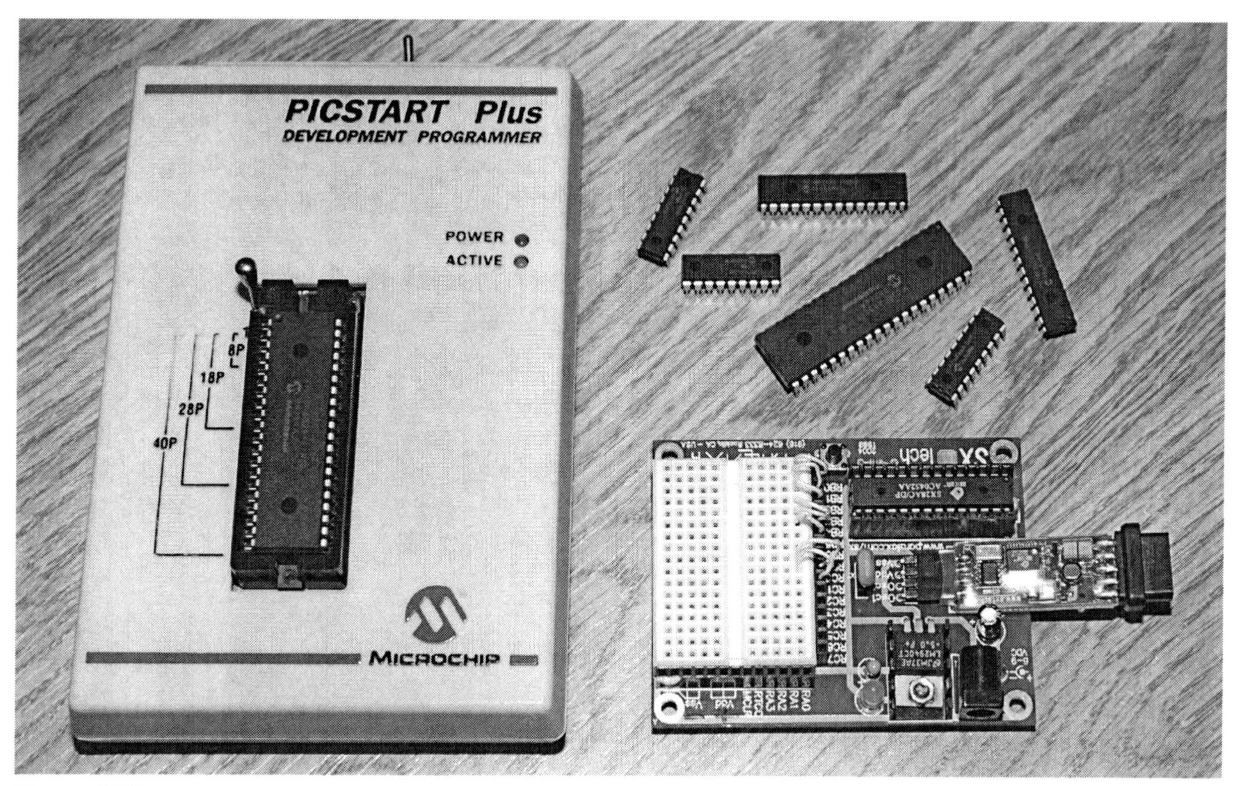

Figure I-27 *Microcontrollers and programmers.*

Beyond the Breadboard

Once you have your prototype up and running on your breadboard, you will need to move the components to a more permanent home so that your device can fit into a cabinet and become real hardware. This migration usually involves moving everything to a fiberglass board with holes with or without copper traces and then adding the wires. This perforated board is usually called *perf board* and is available in many sizes with or without copper-plated holes or strips that copy the connections on a solderless breadboard. Figure I-28 shows an empty and populated copper-hole perf board with a dual stepper-motor drive prototype soldered and tested.

For small circuits, I usually use a non-copper-plated perf board and just solder wires on the bottom of the board where the component leads stick through because this is easy, quick, and inexpensive. Larger circuits become a mess quickly, so having copper-plated holes or even traces like a solderless breadboard can make migration a lot easier. Perf board is available as a large sheet, and it is easy to just cut off a piece the size you need using a utility knife to score the surface and then break it apart. A perf board circuit made correctly is just as good as a manufactured PCB, so don't underestimate what can be done with a handful of wires, a few dollars worth of perf board, and some hard work. Have a look at the complex VGA computer prototype on the huge breadboard in the top half of Figure I-29. The prototype now lives on an 8 × 5 in perf board (lower half of the figure) and functions perfectly at 20 MHz.

Figure I-28 *A perf board becomes a permanent home.*

Figure I-29 *My Retro Game System on a massive breadboard.*

The next step after hardwiring your project to a perf board will be an actual PCB like the one shown in Figure I-30. When you are using surface-mounted components with very small pitch leads, you don't have many options besides a real PCB. A PCB also makes more sense when you need to duplicate a project or want to get into selling your designs. Bringing a professional PCD to market is a costly and time-consuming process often involving many failed attempts and board revisions, but for us hobbyists, there are alternatives that may be viable if a real PCB is needed. If you search the Internet using "PCB production," you will find several companies that will allow you to download their free software to design your PCB and then order a few boards for around $100. This is actually a fair deal considering the amount of investment and time it would take to produce your own PCB at home using a chemical etching– or photo-based system. Of course, you will have to decide if the cost is worth it.

Well, let's dig into some of the projects and have some fun. Remember that we all had to start from the beginning, and if something does not make sense at first, turn off the power supply, take a break, and search the Internet for more information. Electronics is a great hobby, and with a little investment in time and money, you will be able to create just about anything. When you start to get a good handle on things, feel free to try out your own modifications to the projects presented in this book, or mix and match them together to create completely new devices. Oh, and don't feel too bad when you accidentally let the smoke out of your semiconductors—they won't feel a thing!

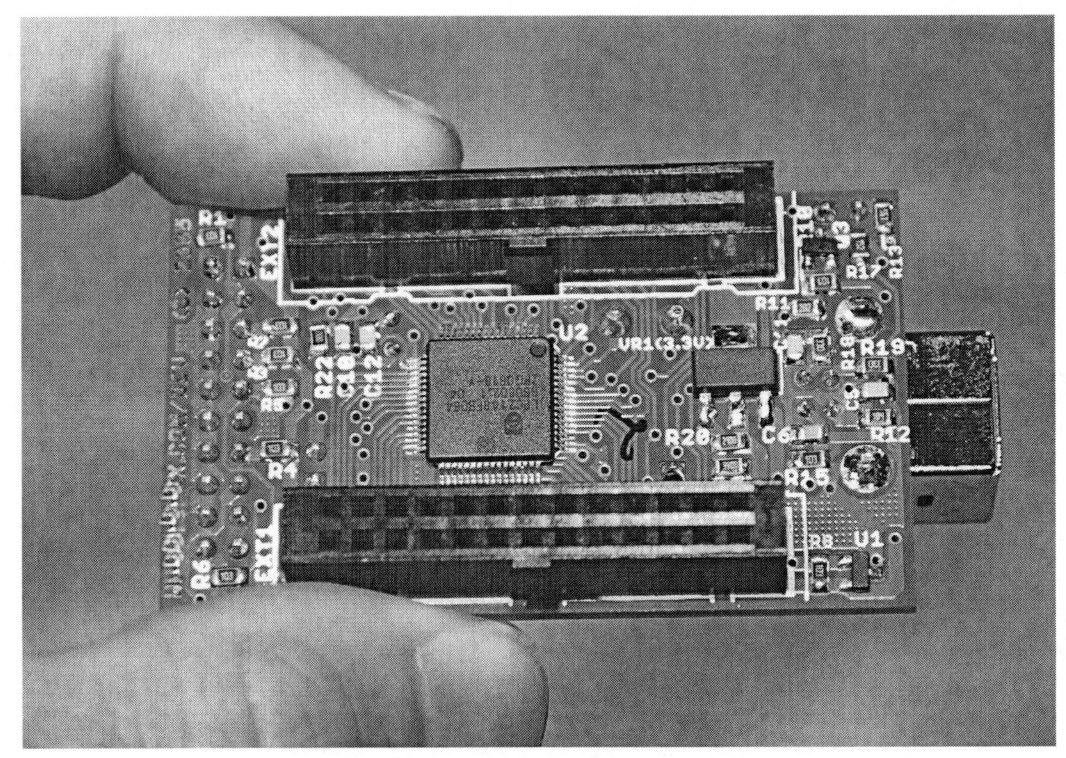

Figure 1-30 *A professionally made printed circuit board.*

Section One

Exploring the Human Body

This section is about measuring and testing reactions that occur in the physical body. Although the mind is not in complete control of some of the involuntary physical functions, a certain level of control can be measured and enhanced with proper training using feedback. Heart rate, body temperature, and even reaction time can be controlled to a certain degree with a little training and with the help of some of the devices presented here. Maybe you want to enhance your physical state before an athletic competition? How about showing your friends how you can "beat" a lie-detector test (polygraph)? By training your mind to react based on the output of some of these devices, you can train and enhance the mind and body link and become more in tune with your own physical machine.

Project One

Biofeedback Device

This biofeedback device measures the electrical resistance of the skin and then changes the tone of an audio oscillator depending on the reading. This type of body measurement is also known as *galvanic skin response* (GSR), *electrodermal response* (EDR), *psychogalvanic reflex* (PGR), and *skin conductance response* (SCR). This type of device has been used in the medical field to measure a patient's emotional response and to treat disorders such as phobias, anxiety, and stuttering.

One interesting and controversial use of a biofeedback device is called an *E-meter*, which is used in some forms of the Dianetics and Scientology auditing. This device is formally known as the *Hubbard electrometer*, for the Church of Scientology's founder, L. Ron Hubbard. Interestingly, the Church of Scientology restricts use of the E-meter to trained professionals, seeing it as a religious artifact that can measure the state of electrical characteristics of the static field surrounding the body. The meter is believed to reflect or indicate whether or not the confessing person has been relieved from the spiritual impediment of his or her sins. It can be used only by Scientology ministers or ministers in training, and these devices are manufactured at the Church of Scientology's Golden Era Productions facility in California.

Biofeedback also can be used to measure a person's response to physical activity because the direct result of exertion will be a response in the sweat glands. Maybe you need to learn to speak publicly without breaking a sweat or to beat a lie-detector test? No matter what your "evil genius" motives are, you probably will find ways to use the biofeedback device for your own agenda.

The biofeedback device is a voltage-controlled audio oscillator that increases its frequency as resistance decreases. Thus, the more you sweat, the higher is the pitch of the resulting output. The oscillator also has a volume control so that you won't go insane from the nonstop sound that it produces while in use. For silent operation, the speaker can be removed and the output fed into any multimeter with a frequency-measuring function to display the results in hertz (Hz) rather than an audio signal. Let's review Figure 1-1 to see how the biofeedback device works.

Transistors Q1 and Q2, along with R1, R2, R3, R4, C1, and C2, form a basic audio oscillator that runs on a 9-V battery. To make the tone of the oscillator change in response to voltage, Q3 acts as an amplifier that feeds a voltage back into the circuit between R2 and R3, changing the output frequency. Since the base of Q3 is connected directly to the subject's body along with the 9-V

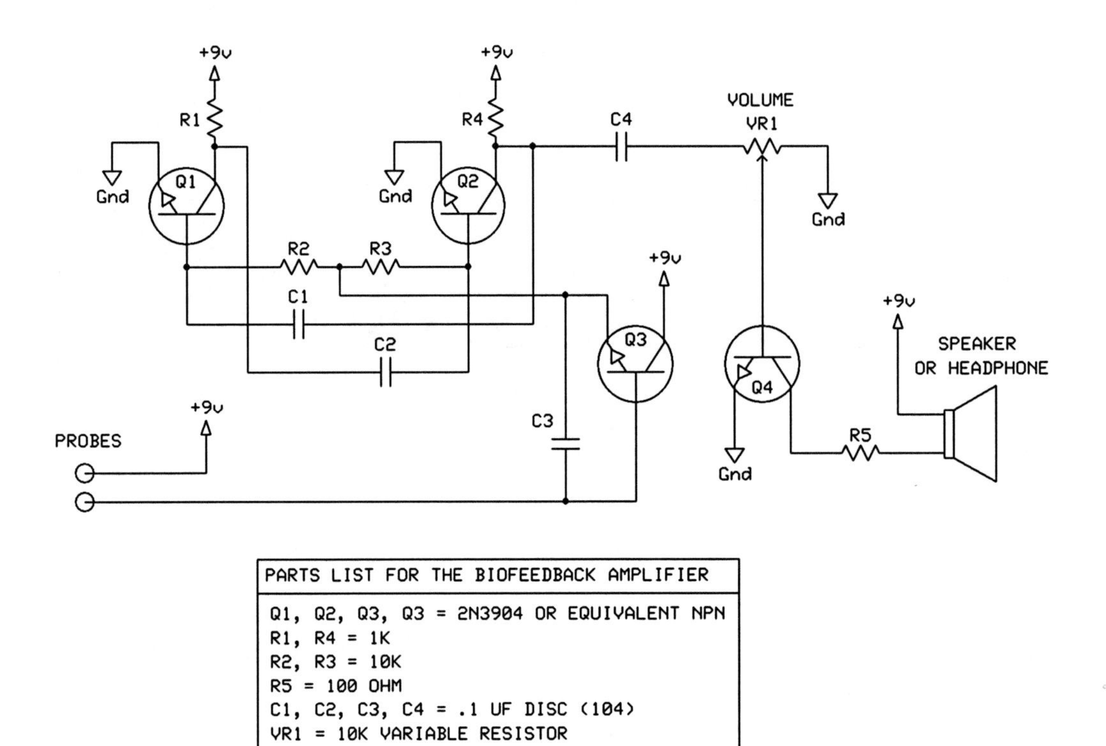

Figure 1-1 *Biofeedback device schematic.*

supply, the resistance of the skin creates changes in the voltage feed into the oscillator. To give the audio oscillator some output gain, Q4 is a simple amplifier that can be controlled by turning the variable resistor VR1.

A piezo buzzer also can be used in place of a speaker in case you are tight for cabinet space or don't want the full volume of a speaker. The schematic shown in Figure 1-2 feeds the output from amplifier transistor Q4 through a piezo element rather than to a standard speaker. Piezo buzzers are small coin-sized disks with a bit of crystal attached to one side that respond to voltage changes. These piezo buzzers or elements can be found in cordless phone bases, handsets, and many other devices that beep or blip. The cover of any digital watch is also a piezo buzzer. Figure 1-3 shows a few speakers taken from my junk box along with a piezo buzzer in a plastic case (*bottom right*). Piezo buzzers can be as small as a thumbtack and produce a decent level of sound, so they are great when you need to save space in a black box.

As for the speaker, any large or small speaker with a rating of 4 to 16 Ω will work fine with this circuit, although you probably don't need anything too large. A speaker taken from a dead portable radio would be good. It's always a good idea to breadboard a project before considering soldering the components in case you want to make modifications or test a component

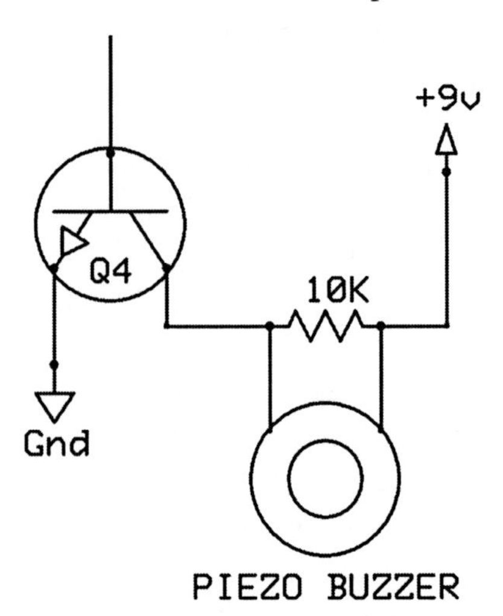

Figure 1-2 *Piezo buzzer alternative output.*

Figure 1-3 *Possible output speaker selections.*

substitution. This circuit is pretty forgiving, so practically any small NPN transistor probably will work. When you are initially testing the breadboard version, just grab hold of two wires for probes because that will work just fine. Figure 1-4 shows the biofeedback circuit being tested on a solderless breadboard.

Once your circuit is working properly, you should hear nothing out of the speaker until you grab hold of the probe wires or short them directly together. When shorted together, the audio oscillator will generate its highest frequency, which will be similar to the output if a person is really sweating profusely. If you do not hear any audio when the probe wires are shorted together, turn the variable resistor VR1 back and forth to make sure that the volume is not turned off. If there is still no output, then you have either a wiring problem or a wrong component. After 30 years as an electronics hobbyist, I have learned that 99 percent of all problems are wiring problems, so recheck the wiring if your circuit fails to work. Once you are happy with operation of the circuit, it can be moved to a more permanent home for installation into a cabinet.

Unless my circuit has a large number of components or integrated circuits (ICs), I always snap off an appropriately sized piece of perf board and then add the wiring on the underside of the board. Copper-clad perf board is also nice, but it can cost a lot more and make it more difficult to change or fix wiring at a later date. Since there are only 13 small semiconductors in the biofeedback device, it fits nicely on a 1 × 2 in piece of perf board, as shown in Figure 1-5. If you copy the general parts layout that was used on your breadboard, then it will be easier to add the wiring. Also, if you have the spare parts, it is a good idea to leave the breadboard version alive while you build the perf board version so that you can compare the two if something does not work on the final version.

Most of the simple devices in this book will fit into a small plastic hobby project box with room for a battery and necessary switches. Most electronics supply stores will have various sizes

Figure 1-4 *Testing the circuit on a breadboard.*

Figure 1-5 *Migration from breadboard to perf board.*

of plastic boxes, and you could even use those gray ABS plastic electrical boxes found at hardware stores to mount your projects. If you are on a really tight budget, you always can get creative and look around the house for a suitable enclosure, such as a soup can or an empty product container. Figure 1-6 shows the battery, variable resistor, on/off switch, and mounting terminals fit into a small black plastic box purchased from RadioShack. I always install an on/off switch between one of the battery wires by default, and a circuit like this can run from a good battery for a long time.

Once you find a project box large enough to hold the battery and hardware, you will need to consider the size and shape of the circuit board. Figure 1-7 shows how I came up with the 2 × 1 in size for the perf board, which was just large enough to fit alongside the battery and contain the 13 components. If the underside of the circuit board is in contact with the battery case or any of the mechanical parts, you can wrap it in electrical tape once it is tested and working to create an insulating barrier. Also, don't forget about all the connecting wires when choosing a cabinet because they can take up a bit of space when it comes time to cram the lid on the box. Now, the biofeedback device is almost ready for use.

Although the biofeedback device works perfectly fine just by grabbing the probe wires between your fingers, this will not be a reliable way to measure your skin resistance because the harder you grip the wires, the more resistance you create. Someone trying to "trick" the unit simply could vary the tone by changing his or her grip on the probes, so some way to attach them to the body will be necessary. Two methods I have found that work very well are copper finger bands and copper plates taped to the subjects' arms. Both finger bands and plates can be made from some copper tubing from the hardware store, as shown in Figure 1-8. The finger bands are small lengths of 1-inch-diameter copper tubing with a slit cut along the length so that they can be safely placed over a finger. The slit allows the diameter to be tweaked, if necessary, by placing a flat-head screwdriver in the slit to widen the opening. This is also a safety release function in case the ring becomes stuck on a finger. The flat plates shown in Figure 1-8 are also bits of the same copper

Figure 1-6 *Finding a suitable cabinet.*

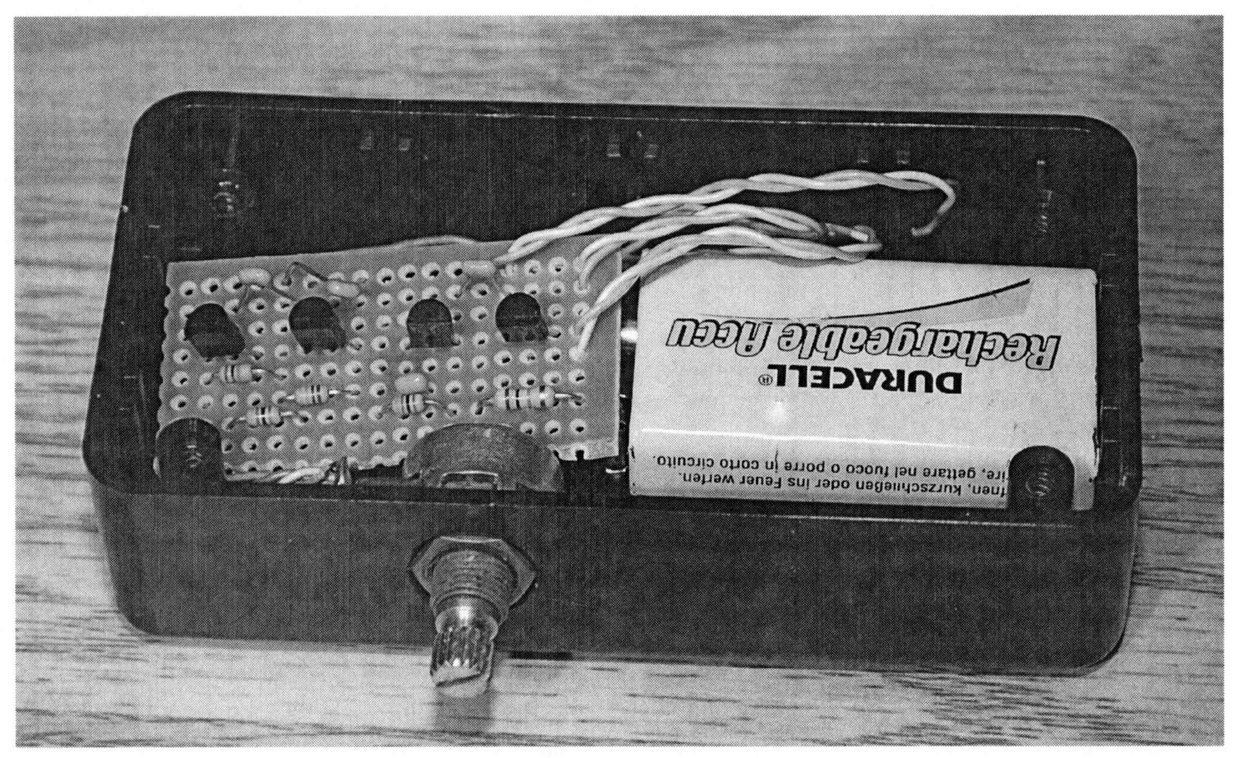

Figure 1-7 *Test-fitting the circuit board.*

Figure 1-8 *Making finger and body probes.*

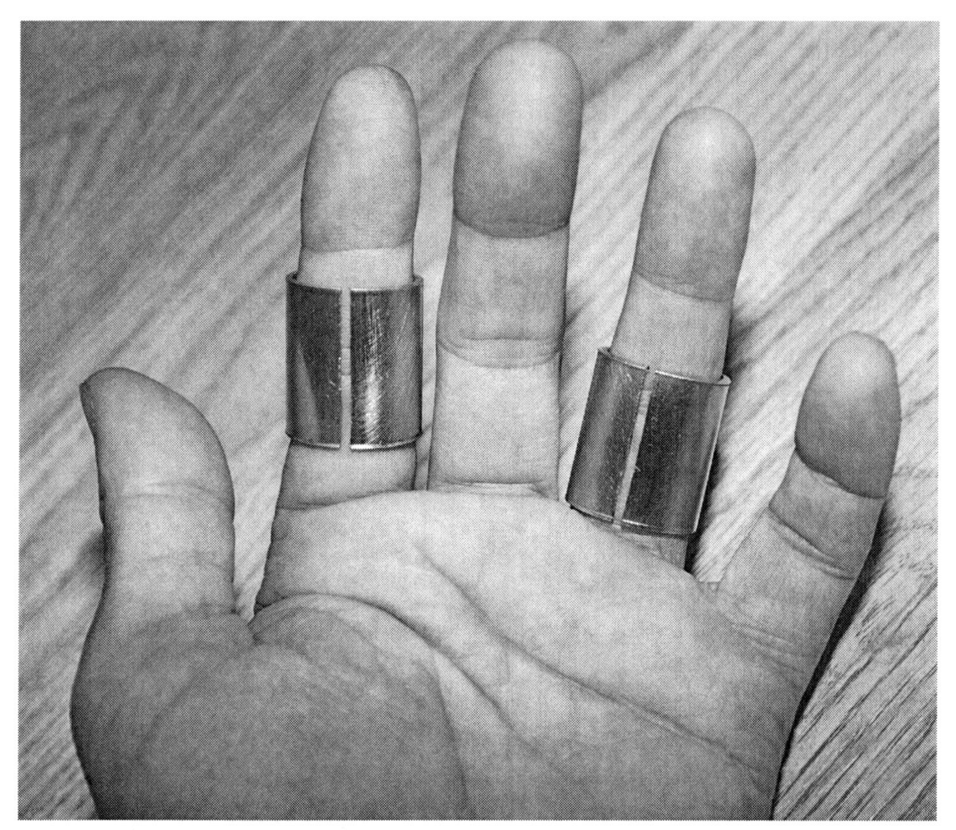

Figure 1-9 *Test-fitting the two finger probes.*

tubing hammered flat into plates. All edges should be sanded or filed so that there are no sharp edges, and the copper can be cleaned with some steel wool for best conductivity.

The 1-inch-diameter copper tubing should be a good fit for most people, either a finger or thumb. If you find that the rings are too small or too large, then adjust them by squeezing or prying open the slit to change the diameter. The rings should slide on easily but also make good contact for the biofeedback machine. Some type of probe jelly normally is used to make a better skin connection, but in this case that would defeat the purpose of the machine because it is the resistance between the skin and the probe that is being measured. For this reason, probes are always installed on dry skin.

The quick-disconnect jacks shown in Figure 1-10 are useful because you can easily change the probes when needed. You might even want to try multiple probes in different series or parallel configurations to see what happens. Now that your biofeedback device is ready to use, you must come up with some creative "evil genius" ways to use it. To verify that it is functioning properly, get comfortable, and connect the probes so that you get a steady unchanging tone out of the box. It may take a minute for the tone to stabilize as the moisture between the probes and your skin settles, so relax until the frequency seems steady. Once the tone has not changed, contract your leg muscle as hard as you can so that you exert yourself somewhat, and the tone should increase slowly as your skin resistance changes. Responses will be slow and gradual, with increases in frequency occurring much faster than decreases because moisture evaporates to lower the frequency. For more accurate and quiet readings, you can connect a frequency meter across the speaker output or remove the speaker completely and just measure the frequency.

Figure 1-10 *The completed biofeedback device and probes.*

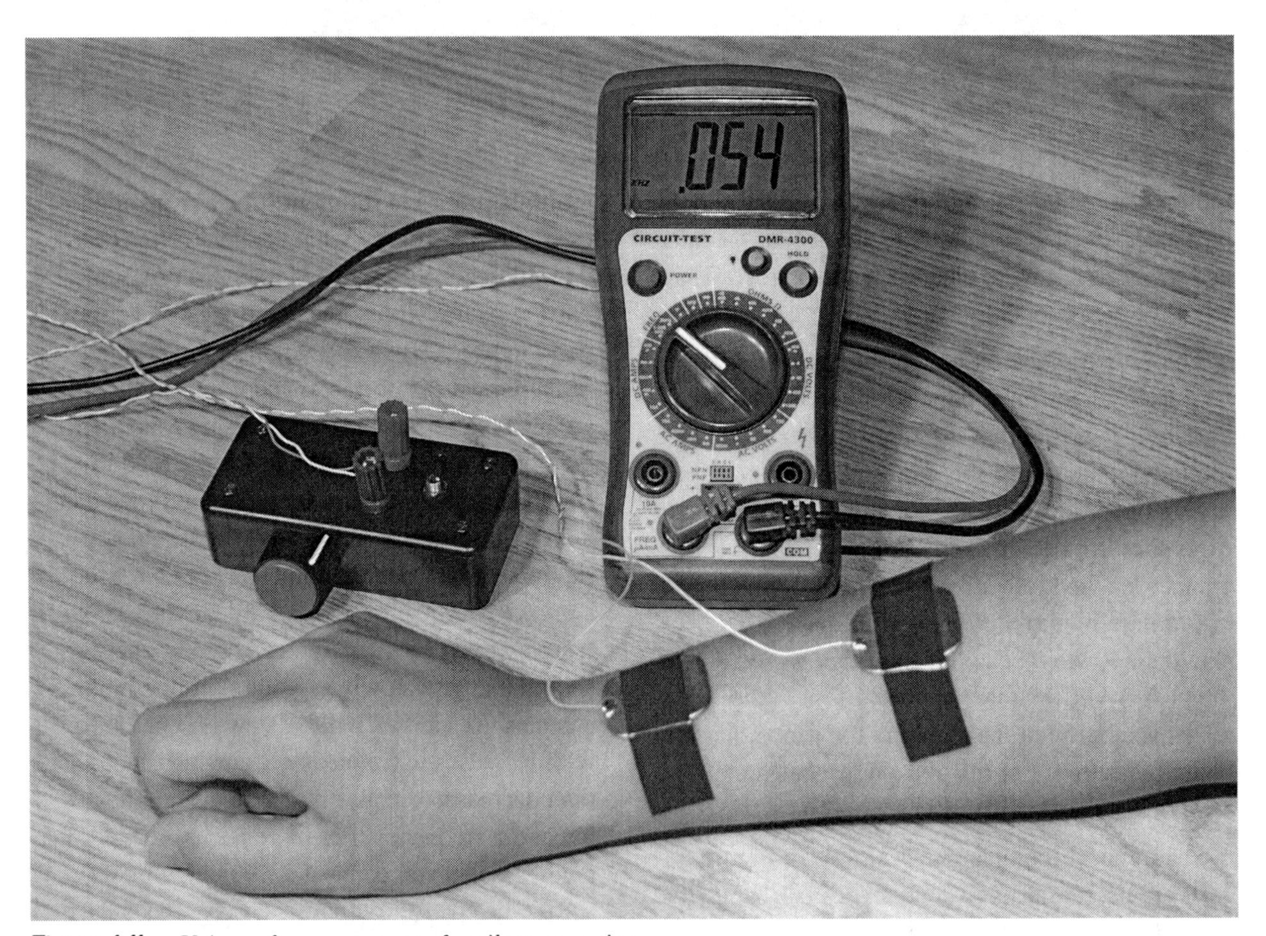

Figure 1-11 *Using a frequency meter for silent operation.*

I added an output jack to my cabinet so that I could plug in an external frequency meter, as shown in Figure 1-11, for more accurate and silent operation of the device. Silent operation may be preferred if you don't want your subject to influence or try to trick the machine. By connecting a frequency meter instead of a speaker, you can see much smaller changes in the frequency, which also may help those who are a little tone deaf when looking for small changes. My multimeter also has a serial port, so I can connect it to a PC and graph the results in real time, making it easy to compare a long-term test over time. Also notice in the figure the use of arm probes, which are held in place by some electrical tape. You may have to play around with the best place to connect the probes for the desired result, but keep in mind that bare skin makes for better testing.

Well, I hope that you have fun with this device and that there are many other things you can do to modify or improve on the design. The placement and types of probes used certainly will affect the results, so get creative and try some now ideas. Maybe a pair of metal spheres held under the armpits? How about some kind of forehead band? Another thing you could try doing is to use the device to check moisture in soil by making probes out of some nails. Add a power supply for continuous operation, and you now have a water flood alarm for your basement by placing the probes in a problem area. I'm sure that any "evil genius" will come up with all kinds of uses and modifications for this device. Have fun!

Project Two

Reaction Speedometer

This reaction speedometer will measure a person's motor reaction time to a series of lights or a trigger sound. The test begins as soon as the tester flips the switch on the main unit, causing the 10 light-emitting diodes (LEDs) on the box to begin lighting in sequence. The subject is instructed to flip his or her handheld switch as soon as he or she sees either the first LED light or when he or she hears a sound from the optional sound add-on circuit. The fewer LEDs that are on once the subject flips the switch, the higher is his or her reaction time. To increase the difficulty of the test, a variable control allows the LED sequence to be adjusted from a slow crawl to a lightning-fast chase that few will be able to keep up with.

Because this project also includes an optional sound add-on that will send an audio tone once the first few LEDs light up, you can test your subject's motor responses to light, sound, or both at the same time. The reaction speedometer also can be used to hone one's response time for such things as improving in sports, martial arts, or even video games. Daily testing also can give some cues to alertness or the effects of such things as sleep or caffeinated beverages.

The heart of the reaction speedometer is the 74HC4017 (or 74LS4017) decade-counter IC, which can turn on one of 10 outputs in sequence each time a clock pulse is sent. The clock pulses are variable, so the speed of the LED sequence can be controlled to make the test easier or harder.

Figure 2-1 shows the main part of the reaction speedometer schematic without the optional sound output circuit. As you can see, the 4017 decade counter (IC2) is in charge of lighting all the 10 LEDs in sequence, which will happen at a rate controlled by the clock-pulse circuit made from the 555 timer (IC1). There are also two switches—one that allows the tester to start or reset the test sequence and another to allow the subject to freeze the sequence in order to complete the test. A variable resistor (VR1) adjusts the speed of the clock pulses so that the test can be adjusted to the subject's best abilities.

Although the schematic is quite basic, it is always a good idea to first build the device on a solderless breadboard for testing so that you can verify its operation and make any possible modifications before heating up the soldering iron. You also may want to build the optional sound add-on to enhance the test with an audio cue as well as the visual light show.

The optional sound add-on lets the subject respond to an audio tone that will occur as the first few LEDs light up, enhancing the test. The simple schematic shown in Figure 2-2 is an

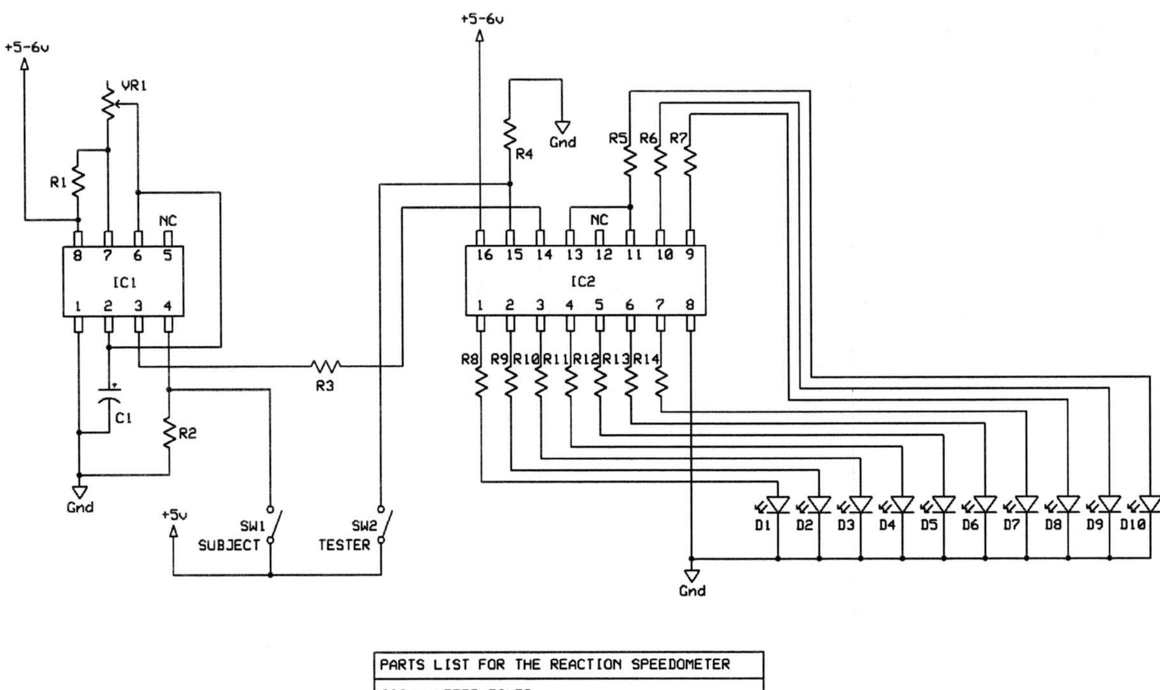

Figure 2-1 *Reaction speedometer schematic.*

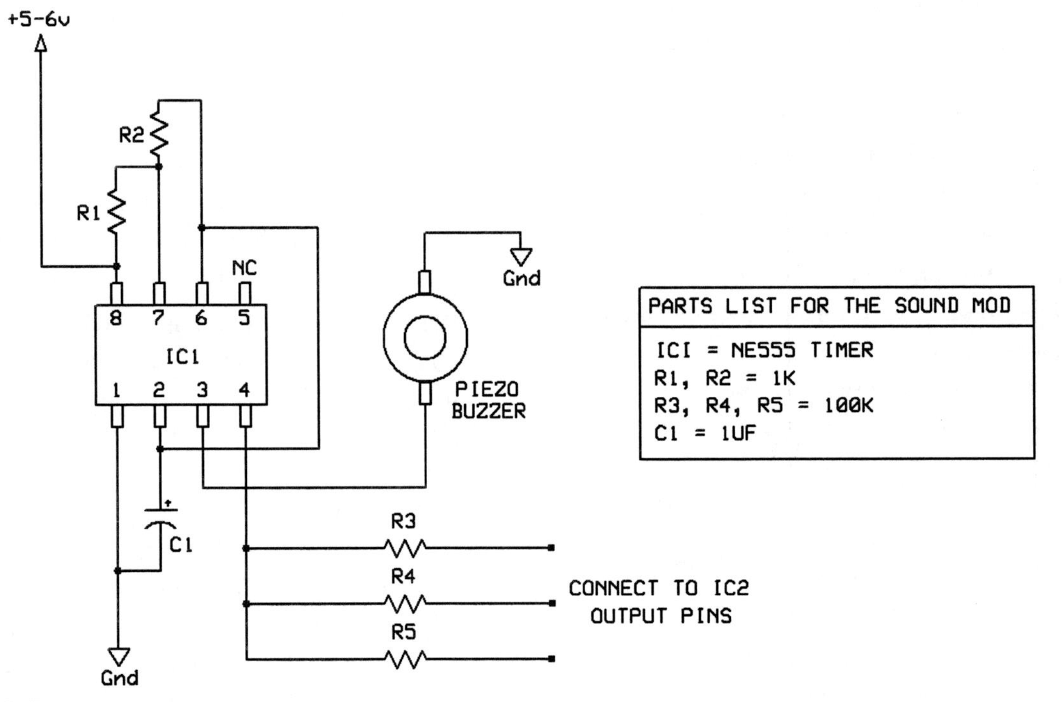

Figure 2-2 *Sound add-on schematic.*

audiofrequency oscillator made from another 555 timer and is practically the same circuit as the original 555 clock circuit shown in the main diagram. The output from the timer is fed into a piezo buzzer so that an audible tone will be heard as soon as the circuit is triggered. Triggering is accomplished via pin 4 and the resistors (R3–R5) that connect to the output pins on the 4017 decade counter.

Depending on which output pins you connect to the sound trigger, you can alter the length and timing of the audio tone as the 4017 steps through its outputs. In my reaction speedometer, I decided to connect to output pins 2, 3, and 4 so that the sound is on for about a third of the time as soon as the test sequence begins. I did not connect to the first output pin because doing so would mean that the sound would be on while the test is in ready (reset) mode, which causes the first LED to remain lit. If you want the tone to last longer, add more connection points. If you want the tone to start later on in the sequence, move the connection points to higher output pins on the 4017 counter. To alter the tone, play around with different values for R1, R2, and C1.

Figure 2-3 shows the reaction speedometer built on a solderless breadboard for initial testing. Notice how each of the three subcircuits (i.e., sound, clock, and counter) are shown in the figure. To make the circuit work, you also will need a pair of single-pole, single-throw toggle switches for the subject and tester. The tester's switch will be mounted to the main box, or it also could be made remote like the subject's switch, to be held in one hand. In one position, the tester's switch will reset the counter, causing the first LED to stay lit. In the other position, the test will start, causing every other LED to light in sequence at a rate controlled by the variable resistor in the 555 clock circuit.

Once the test is in motion, the subject must throw the switch as fast as possible to freeze the test, which will cause the clock to stop. If the test gets all the way to the last LED, the subject has failed, and the test will need to be reset. If the subject cannot beat the test before the last LED is lit, the tester will have to lower the clock rate until the subject can freeze the test somewhere between LED number 2 and LED number 9.

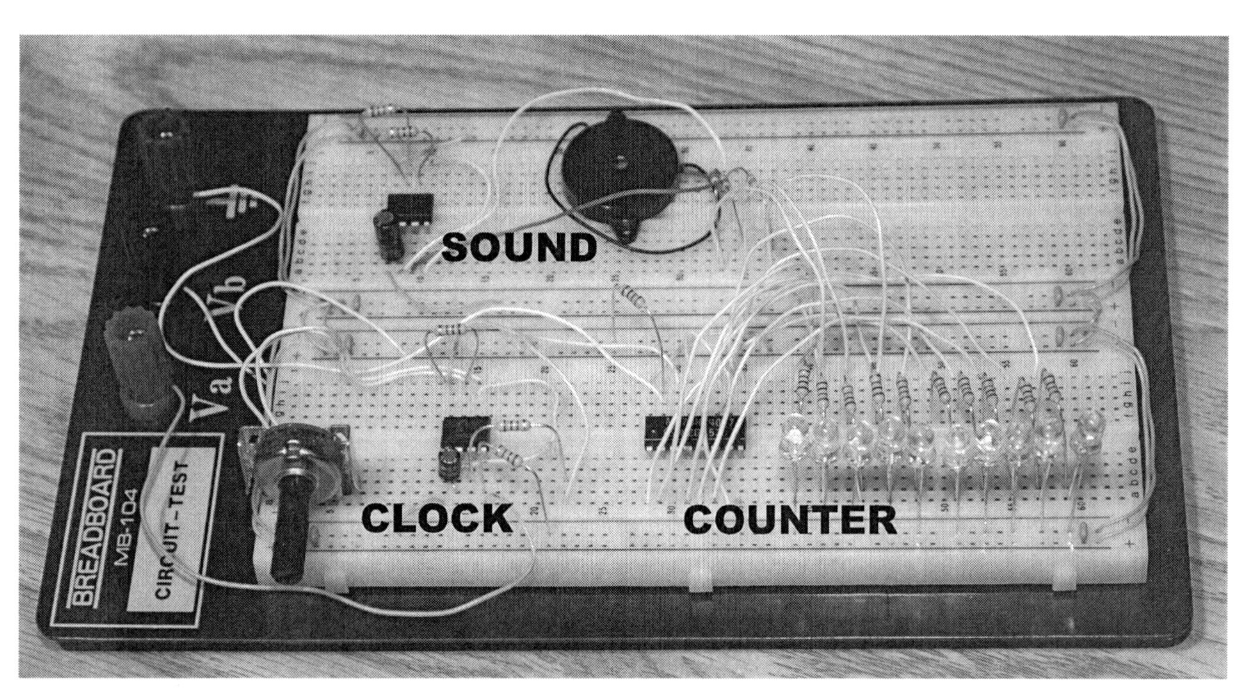

Figure 2-3 *Breadboarded test circuit.*

When you have verified the operation of the circuit and made any necessary modifications, you can move the components to a more permanent home on some perforated board for installation into a cabinet.

Since I added the sound part of the circuit after I built the main circuit, only the clock and counter are on the perforated board shown in Figure 2-4. There probably was enough room to jam the sound add-on onto the board as well, but I thought it might be handy to have the sound board separate to be reused as a generic audio oscillator in some other project later. In my usual style of building a perf board circuit, the component leads are bent on the underside of the board, and all wires are also soldered on the underside of the board. There a quite a few wires coming from the circuit board owing to the 10 LEDs, two switches, variable resistor, and power wires needed.

The sound add-on circuit board shown in Figure 2-5 is completed in the same manner as the original circuit, using a small bit of perf board and adding all the wires to the underside of the board. Also shown in the figure is the small piezo buzzer, which could be replaced by a standard speaker if you wanted by adding a 250-Ω resistor in series with the speaker leads to reduce current draw on the timer output pin. Having the sound circuit on a separate board is handy if you need a simple sound system that can be triggered by a voltage change on some output pin.

There are several ways that you can set up the reaction speedometer for use. The simplest method is to have all the tester's controls in one main box, and place the subject's switch remotely in some type of handheld container. You also could make the tester's start switch the same way so that the LED speedometer is placed between the tester and the subject. A simple handheld switch unit can be made by placing the toggle switch inside a small plastic box or container such as the film container shown in Figure 2-6. The subject now can hold the unit in one hand

Figure 2-4 *Biofeedback circuit on a perf board.*

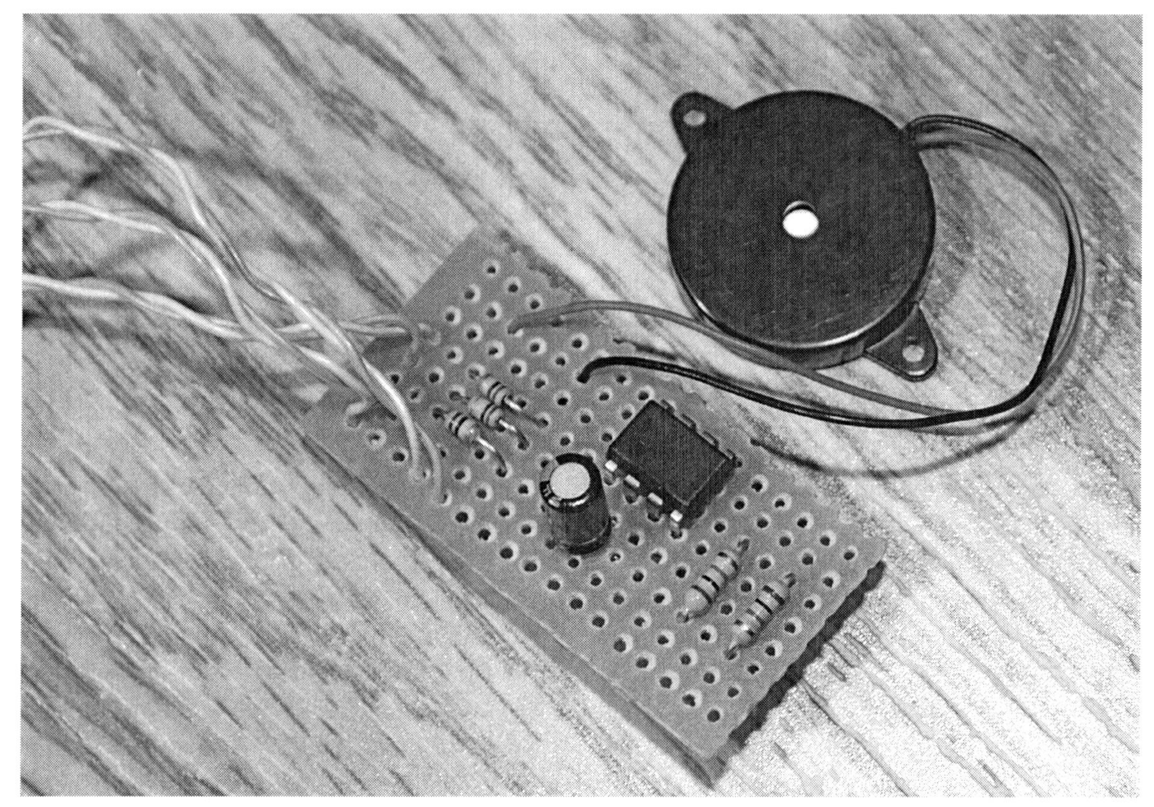

Figure 2-5 *Sound add-on perf board.*

Figure 2-6 *Subject's handheld switch.*

and work the switch with his or her thumb. I kept the tester's switch on the main box, but the advantage to having both switches remote from the main box is that there will be no distractions as the tester flips the switch to start the LED sequence counter.

Both circuits will run from 5 or 6 V, so you can power the unit with either four 1.5-V batteries in series or a larger battery or power pack and a regulator. The four AA batteries shown in Figure 2-7 will give the circuit 6 V and run for a very long time because the unit uses very little power. Although most 74 series logic chips, like the 74HC4017, specify only 5 V, they usually will run with higher voltages, so 6 V is not a problem. For use with a higher-voltage battery, such as a 9-V battery, you certainly will need a regulator to reduce the voltage to avoid damaging the chip.

The simple 5-V regulator schematic shown in Figure 2-8 will allow you to connect a 9- or 12-V battery or a direct-current (dc) wall adapter with a voltage between 9 and 15 V to practically any project in this book requiring 5 V. Because all the circuits in this book use very little power, a heat sink will not be needed because the 7805

Figure 2-7 *Six volts from four batteries.*

regulator will not be working very hard at all. R1 and D1 are optional, but they do let you know when your project is switched on. Having an ON-indicator LED is nice, especially if there is no way to tell if your circuit is on when you put it away.

Figure 2-9 shows the simple 5-V regulator made on a small bit of perf board and ready for use with any of the 5-V projects presented in this book. The ON-indicator LED and current-limiting resistor are not shown on the board because they usually are placed somewhere on the cabinet front, away from the circuit board. If you have modified your circuit to include many more components, then your regulator may need a heat sink to dissipate the extra heat away from the case. If the regulator is so hot that you can't keep your finger on the small metal tab, then you should add some type of heat sink to help cool the device. Any small bit of steel or copper plate usually will be adequate to cool the regulator. If your entire circuit uses more than 1 A, then you probably will have to find a larger regulator than the 7805.

To make the reaction speedometer a little more interesting, I made an oval 10-digit speedometer using Photoshop and then printed it out so that I could glue it to the plastic box. This template also served as a guide when drilling the holes for the 10 LEDs. To drill a perfect hole, start with the smallest drill bit in your kit, and your hole will not wander as you use larger bits. As shown in Figure 2-10, the pilot hole (second drilled hole) is a lot smaller than the hole needed for the $^3/_{16}$-inch-diameter LEDs that I am using. It is good practice to drill a small pilot hole anytime you need to make an accurate drill hole in a cabinet. Having multiple holes line up is important when there are a few controls or lights in a row because the accuracy of your work will really show.

The completed reaction speedometer is shown in Figure 2-11, ready to help me train my reflexes to the cutting edge. I can set the speed dial to about half and still pass the test before the last LED lights, but any more than that is just way too

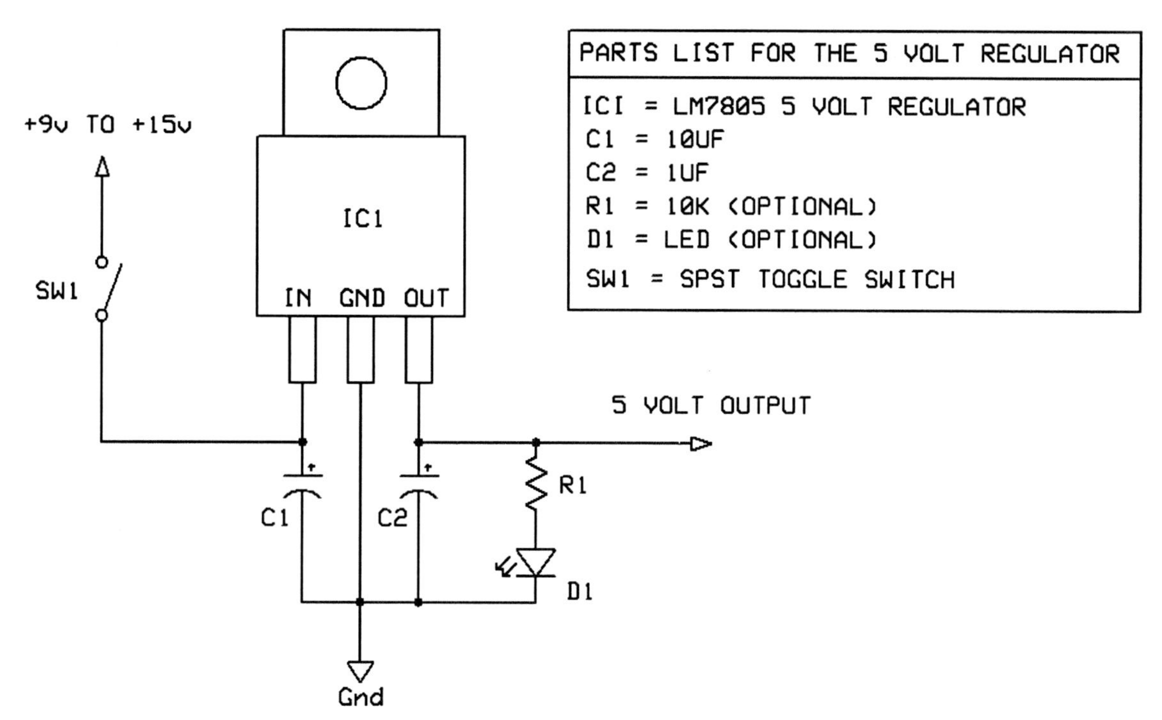

Figure 2-8 *Optional regulator schematic.*

Figure 2-9 *Five-volt regulator breadboard.*

Figure 2-10 *Adding a graphic speedometer.*

Figure 2-11 *Reaction speedometer ready to use.*

fast. I have noticed that after drinking a caffeinated coffee, I can flip the switch between one and two LEDs quicker then when I am not buzzing from caffeine. When I am tired, I notice that my speed decreased by one or two LEDs. To get the most accurate results, I take the test 10 times, add the results, and then divide by 10 to get the real speed average. Another thing I have noticed is that when using only the sound as a trigger, I am faster than when using only the LEDs. I guess my ears are quicker than my eyes!

The reaction speedometer works very well, and there are a lot of interesting modifications you could make to enhance the design. Adding colored LEDs might make the test look more interesting. How about replacing the piezo buzzer with a headphone jack so that you can have your subject wear headphones? Maybe you could design a random trigger using a few more counters and a clock so that you can test yourself. How about a tactile test where the output from one of the LEDs is fed into a transistor or relay that switches a solenoid placed on the subject's body? I am sure you will find many ways to modify and experiment with this simple circuit in order to satisfy your own "evil genius" needs.

Project Three

Body Temperature Monitor

Body temperature varies throughout the day depending on our mental and physical state. Normally, our bodies remain at a temperature of 98.2°F (36.8°C) when we are awake and not involved in strenuous physical activity. Measuring body temperature during sleep experimentation is particularly useful because our bodies drop to their lowest normal temperatures when we reach the second half of our sleep cycle. This low-temperature point is called the *nadir* and will be about a degree cooler than the normal waking body temperature.

The device presented here is an example of how an inexpensive digital thermometer IC can be connected to a microcontroller to monitor body temperature. Although this project is very simple and displays only a two-digit temperature value, it would be very easy to modify and expand this example to log data, display decimal values, or interface with a computer.

The Maxim DS1621 is just one of many examples of inexpensive and easy-to-use temperature sensors that can be connected to a microcontroller with very minimal effort. Depending on the device, temperature sensors can output data in many different ways. Some examples include serial data, parallel bytes, analog data, and pulse-width modulation. The DS1621 is a two-wire serial interface, so it is well suited to microcontrollers with a low pin count and can be controlled easily by a basic program. As you can see in the body temperature monitor schematic (Figure 3-1), there is not much to it besides the DS1621 thermometer (IC1), an 8-bit Atmega88 microcontroller (IC2), and a pair of seven-segment LED displays. The DS1621 sends its data to the microcontroller every few seconds, and then the data are converted to the nearest degree and displayed on the two LED displays. You could easily modify the code to display the decimal values as well or even change the display to Fahrenheit, although that would require three LED displays if you want to see readings over 99°F. I wanted to keep this example as simple as possible because there are so many different ways you could modify this project.

To keep the number of input/output (IO) pins to a minimum, the two transistors (Q1 and Q2) switch between the dual displays at such a fast rate that they all seem to be on at the same time. This time-sharing trick is how most LED displays work, where there are hundreds or thousands of LEDs to control and only a limited number of connecting wires. There really is no limit to how many seven-segment LEDs you can connect as long as you have sufficient drive current and microprocessor speed and can spare the extra IO pin for each common connection. The

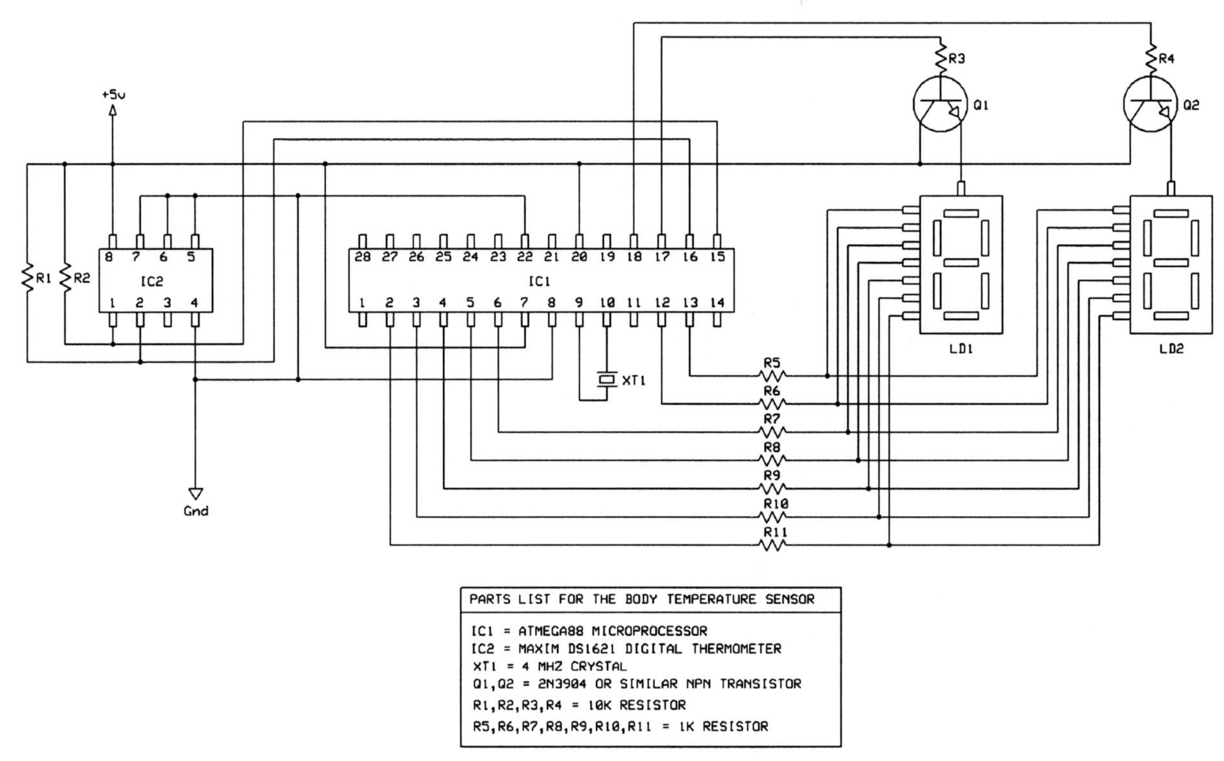

Figure 3-1 *Body temperature monitor schematic.*

seven-segment LED is a very common and inexpensive display that you are already familiar with because it is used in everything from your digital clock to the panel of your microwave oven. There are actually eight segments if you include the decimal point, but we don't use it in this application. These displays come as either stand-alone blocks or chained blocks containing more than a single digit. Some LED displays even have alphanumeric capabilities or multiple "dots" so that they can make any character imaginable. LED displays are either *common cathode* or *common anode*, which means that either the positive connections or the negative connections all go to a common point. It really does not matter which type you use as long as you install them in your circuit so that current is flowing in the proper direction. To use a common-cathode LED display in this circuit, you would have to tie the driver transistors/emitters to ground and connect the common cathode to the collector instead.

Once you have the code compiled and installed in the Atmega88, powering up the breadboard will instantly show you the temperature in your room. Figure 3-2 shows the completed circuit reading 21°C, which is only 1 degree different from what the hallway thermometer was claiming. When I placed my finger on the DS1621, the temperature slowly climbed to 32°C, which seemed about right because I had cold hands, and the temperature outside the body is usually a few degrees colder than inside. For dream research, the actual temperature is not what counts, but rather the variance of temperature over the entire night. Notice the addition of the 7805 regulator on the top right of the breadboard so that the system will run on a 9-V battery.

If you build the circuit compact enough, you may find that it will fit into a cabinet small enough to be placed directly on the body using some kind of elastic strap. I decided to place the electronics in a cabinet but run the temperature

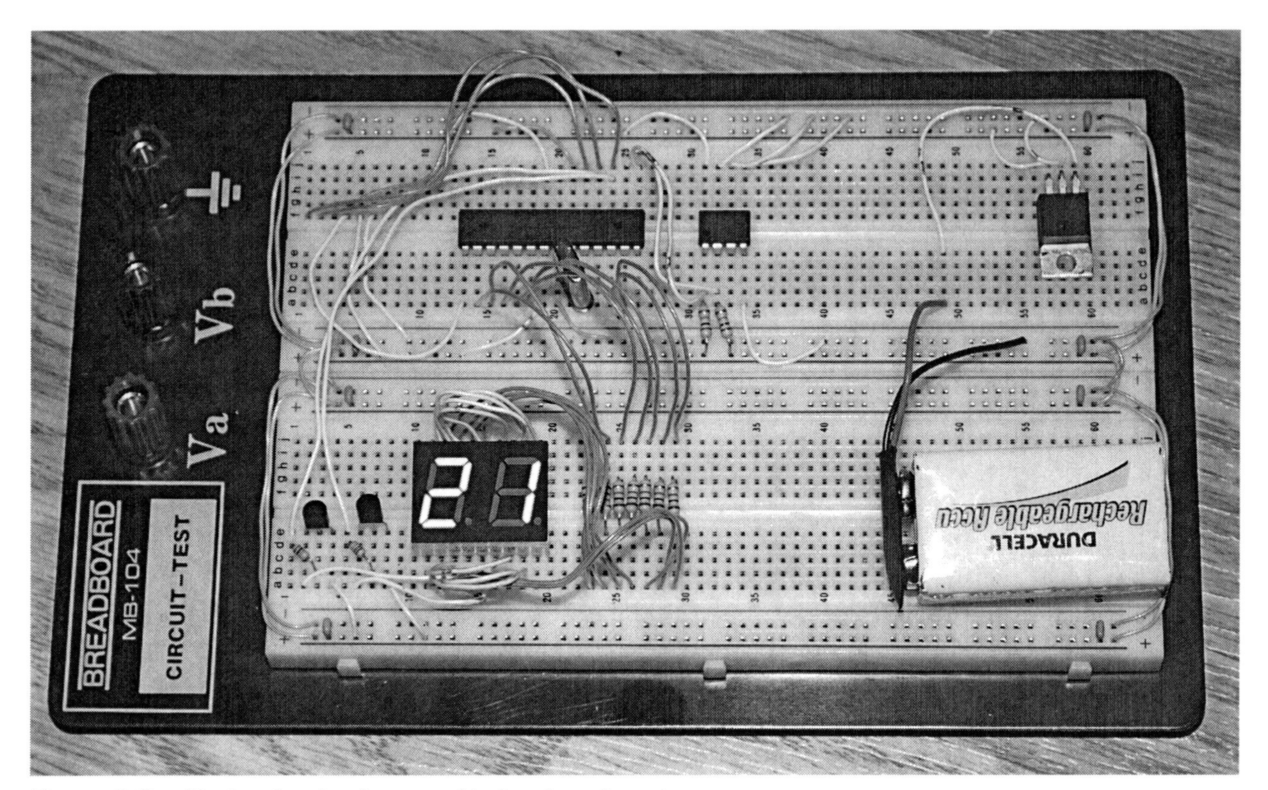

Figure 3-2 *Testing the circuit on a solderless breadboard.*

sensor externally so that it would be more comfortable as a sleep experimentation device. I also wanted a little extra room inside the housing to add some circuitry later that allows a wireless connection through a simple rf transmitter module. Figure 3-3 shows the perf board that carries all the components except for the DS1621 temperature sensor. There will be limits as to how far your temperature sensor can be away from the microcontroller owing to noise and impedance of the wiring, but a few feet should be no problem at all for most devices.

The temperature sensor lives on a tiny bit of perf board with the power and signal wires coming into it (Figure 3-4). I use the top part of a sock to hold the sensor against the subject's arm, where the skin gives good contact with the top of the sensor package. You also might want to run a fine bead of hot glue or nonconductive caulking along the pins of the IC package so that perspiration does not create resistance between the pins of the device if they come in contact with the skin.

The completed body temperature monitor is shown in Figure 3-5, again reading the temperature in my laboratory, which was getting a bit too warm owing to the huge lights I used for making these photographs. Operation of the device could not be more simple; just turn it on and read the temperature, which changes about once per second. I later added one of those microprocessor-compatible rf transmitters to the unit so that I could read the data from the microcontroller into a computer in order to graph the results during the night. Many electronics distributors carry inexpensive rf solutions, which are easy to use and can send serial or parallel data to or from any microcontroller.

The sensor package must come into good contact with the skin to give a reliable reading, and a cut-up sock makes a good armband that keeps the sensor motionlessness and insulates the

Figure 3-3 *The completed temperature monitor on some perf board.*

Figure 3-4 *The remotely located temperature sensor.*

Figure 3-5 *The completed body temperature monitor taking a body temperature reading.*

sensor from the outside world. Different parts of the body will give different temperature readings, so don't be alarmed if your sensor shows that you have a very low body-core temperature. The surrounding hardware also will affect the readings, so you might want to consider adding a line to the source code that compensates for the –3 to –5°C difference this device will have compared with a thermometer stuck under your tongue. If your main goal is reading changes in temperature, then it really makes no difference what the sensor shows so long as it varies along with changes in your subject's body temperature.

The complete source code for the body temperature monitor is shown in Listing 3-1 of the appendix and was written in Bascom AVR to keep it as simple as possible. Basic is a great language for fast prototyping and can be ported easily to any platform in a hurry because it is extremely readable. I will explain what each block of code does so that you can understand the workings of the program and port it to whatever microprocessor you plan to use.

```
' DEFINE TARGET = MEGA88 @ 4MHZ
$regfile = "M88def.dat"
$crystal = 4000000
```

The code following this comment is required to tell Bascom that we are going to target the Atmega88 device and that our clock will be an external crystal resonator running at 4 MHz. Telling the compiler your clock speed becomes important when using commands that deal with timing-sensitive routines such as serial transmission or analog-to-digital readings. Defining the device also helps the compiler to generate user errors that have to do with IO pins. In this way, you can't accidentally try to toggle an IO pin that does not exist on the actual device. Critical timing is not an issue in this program, so you can use whatever crystal you happen to have in your parts box.

```
' CONFIGURE IO PORTS
Config Scl = Portb.2
Config Sda = Portb.1
Config Portd.0 = Output
Config Portd.1 = Output
Config Portd.2 = Output
Config Portd.3 = Output
Config Portd.4 = Output
Config Portd.6 = Output
Config Portd.7 = Output
Config Portb.3 = Output
Config Portb.4 = Output
```

This block of code sets up the pins that will connect to the LED display (outputs) and to the two-wire serial interface to the DS1621. "Scl" and "Sda" are special Bascom reserved keywords that specify the serial data and serial clock lines.

```
' DEFINE VARIABLES
Dim Led(10) As Byte
Dim Msb As Byte
Dim Lsb As Byte
Dim A As Byte
Dim B As Byte
Dim C As Byte
```

Basic uses *variables*, which are letters or words used to hold values. I like to use single letters such as *A*, *B*, and *C* for simple programs such as this one, but when you are working on a large, complex program, use of more descriptive variable names is recommended. "TIMER2" or "REDLED1," for example, would be descriptive variable names that make a lot more sense in a huge block of code. The variables "Msb" and "Lsb" will store the 2 bytes that are returned from the DS1621. "Msb" is the most significant byte and will contain the whole-number value of the current temperature, whereas "Lsb" is the least significant byte containing only the decimal value. "Lsb" is not being used in this version of the code.

```
' DEFINE LED DIGITS
Led(1) = 8
Led(2) = 187
Led(3) = 97
Led(4) = 49
Led(5) = 178
Led(6) = 52
Led(7) = 4
Led(8) = 185
Led(9) = 0
Led(10) = 48
```

This block of code sets up an array of 10 values for the variable "LED." Although this may seem confusing at first, the values correspond to which of the seven segments will be lighted to display a particular number. To make this a little more confusing, the value in parentheses is actually one higher than the represented numerical value, and to light a segment, we want a low bit, not a high bit, so the value of "Led(9) = 0" says that to display the decimal number 8, we want all bits to be off. This means that all segments light up, and an "8" will be displayed. It can be a bit of a chore computing these values by adding port bits, but once you have done it once for your segment, the hard work is completed.

```
' ********** MAIN LOOP **********
Main:
Do
```

Everything from here on is going to happen continually until the word "Loop" is reached, which causes program execution to start again where it first encountered the word "Do." This is called an *endless* loop because it never stops unless forced to by another command or an error.

```
' READ DS1621 TEMPERATURE
I2cstart
I2cwbyte &H90
I2cwbyte &HEE
I2cstop
I2cstart
I2cwbyte &H90
I2cwbyte &HAA
I2cstop
I2cstart
I2cwbyte &H91
I2crbyte Msb , Ack
I2crbyte Lsb , Nack
I2cstop
```

This code block sends the initialization sequence to the DS1621 in two-wire serial format using the built-in "I2C" commands. Having ready-to-use "I2C" basic commands takes a lot of work out of your hands and makes rapid prototyping a snap. The values presented here are only for the DS1621 and are based on the datasheet for the device as well as several code examples found on various forums. The whole-number value is stored in the variable "Msb," and the decimal value is stored in "Lsb" (which is not used). If you are using some other digital thermometer IC, then most likely it will use a different set of commands for communication.

```
' DISPLAY DATA ON LEDS
A = Msb Mod 10
B = Msb \ 10
Gosub Ledshow
```

This small bit of code converts the value stored in the variable "Msb" into a pair of bytes ("A" and "B") containing the two digits that make up the value. This is necessary because each LED display can display only the values from 0 to 9. Once variables "A" and "B" are set, the "Ledshow" routine is called.

```
' RESTART MAIN LOOP
Loop
End
```

The "Loop" command causes program execution to jump back to the main routine where the "Do" command was encountered. This is the end of the infinite loop, and all other commands beyond here must be called by either "Goto" or "Gosub" commands.

```
****** LED DISPLAY ROUTINE ******
Ledshow:
Portb.4 = 0
Portb.3 = 1
C = A + 1
Portd = Led(c)
Waitms 2
Portb.4 = 1
Portb.3 = 0
C = B + 1
Portd = Led(c)
Waitms 2
Return
```

This is the routine that displays a digit on each of the two LED displays. A little trick called *persistence of vision* is used here to switch between displays so fast that your eyes think they are all on at the same time. As you can see, only one bit of the two ”Portb” pins is on at a single time, and then the variables “A” and “B” are sent to the display for only 2 ms each. Because 2 ms is so fast, it appears that each display is on all the time, and the seven-segment lines can be shared, saving valuable IO overhead. Since Bascom array variables start at 1 not 0, the line “C = A + 1” adds 1 to the array pointer so that the value in “Led(c)” is the same as the decimal value we want. This conversion just makes it easier to understand the code, especially when trying to compute the segment bits from scratch the first time. Once the displays have been lighted for 2 ms each, this routine just “returns” to where it was originally called.

That’s all there is to it! A lot can be accomplished in very few lines using Basic, and since microcontrollers work at nanosecond speeds, you have a lot of power at your fingertips.

Since this is a very plain and simple temperature monitor, you likely will want to modify the code to expand and add your own features, such as a decimal readout and maybe a serial or USB interface to send the data to a computer logging program. Simply by expanding the number of LED displays to four, you could send the type bytes as a more accurate decimal value. Adding a wireless link to send data to a logging program for sleep research or dream-state detection is another interesting modification I plan to do later to this simple device. Using body temperature in conjunction with some of the other sleep-research tools really can increase your ability to monitor and predict your sleep and dream cycle through the night. Next, we will build a simple device to measure respiration.

Project Four

Respiratory Monitor

Respiratory rate varies greatly depending on how much oxygen the body requires. The average person will take 10 to 20 breaths per minute while at rest and between 30 and 40 breaths per minute during strenuous physical activity. During sleep, breathing rate also increases, so adding a respiratory monitor to your arsenal of sleep-research tools would be handy. This project presents a novel method of monitoring a subject's breathing rate that uses the noise picked up by a sensitive microphone and preamplifier to feed a recording device. The resulting waveform is very easy to analyze because it will contain visible bursts of data each time the subject inhales or exhales. Any computer program capable of displaying a waveform along with the time can be used to determine breaths per minute or breaths per hour.

The respiratory monitor is actually an extremely sensitive audio preamplifier that exploits the fact that moving air close to a microphone causes an overload in the audio output. It is this high-gain noise that creates the visible bursts of data that can be seen on the computer screen while viewing the waveform. Exhaling creates the largest spike in data, so you can easily determine how many breaths per minute the data contain by either counting the larger spikes or counting all data spikes and then dividing them by two. Figure 4-1 shows the ultra-high-gain preamplifier, which is fed by an electret microphone. An *electret microphone* is a tiny metal can that contains not only a microphone inside but also a sensitive transistorized amplifier so that the output is already amplified somewhat before it reaches the LM358 op amp (IC1).

An electret microphone is the most common type of microphone you will find in audio recording devices and most computer microphones. Even those large plastic multimedia microphones you can purchase for your computer may contain nothing more than a pair of tiny electret microphones inside, along with a chunk of metal to make the device feel heavy (seriously). Answering machines, telephones, and most other small recording devices all will contain an electret microphone, and you also can purchase them new at most electronics suppliers for a few dollars each. Figure 4-2 shows a few of the many electret microphones that I have collected by ripping apart old electronic devices over the years. Some have a rubber cap or may be sealed in plastic, but inside, you will find the same basic tiny metal can with a hole at one end and a pair of leads or solder spots at the other end. All you need to know is which lead is positive and which is negative, but this is very easy to determine.

PARTS LIST FOR THE RESPIRATORY MONITOR

IC1 = LM358 DUAL OP AMP
R1,R3,R4 = 10K
R2 = 1K
C1 = .1 UF CERAMIC
C2 = 4.7 UF ELECTROLYTIC
VR1 = 500K VARIABLE RESISTOR

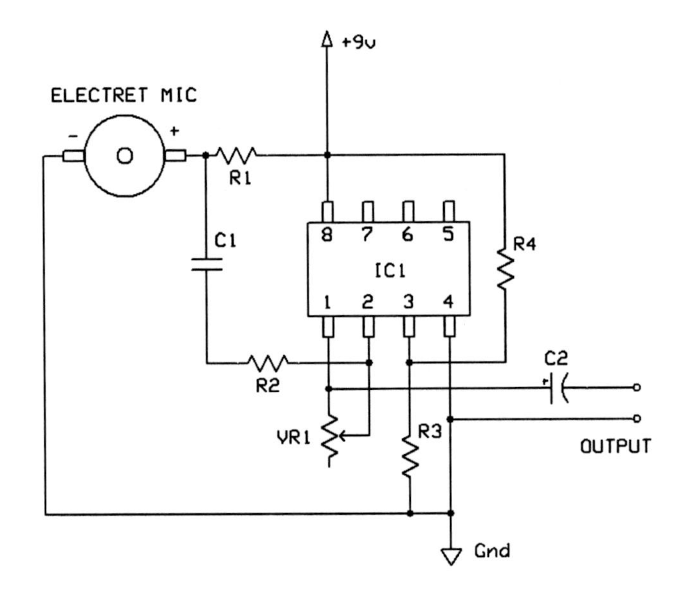

Figure 4-1 *The respiratory monitor schematic.*

Figure 4-2 *Several electret microphones.*

Figure 4-3 *Electret microphones are polarized.*

An electret microphone requires power to run the small amplifier contained within the tiny metal can. Although the power requirement is very small, you still have to figure out which lead is positive and which one is ground so that you can insert the microphone into your circuit correctly. Looking at the underside of the microphone (Figure 4-3), you usually will see that one lead or solder spot is also connected to the metal can by a small trace at the edge. The side that connects to the can almost always will be the negative, or the ground, lead. The good news is that if there is no visual indication of polarity on your microphone, you simply can try it both ways in the circuit without damaging the microphone in any way. If you have the polarity reversed, you simply will get no output from the microphone.

Since the goal is to position the microphone as close to your subject's nostrils as possible, an inexpensive multimedia headset such as the one shown in Figure 4-4 would be perfect for this project because under that block of foam is just another electret microphone. By using a premade headset microphone, you don't have to take the housing apart because you can simply add a $\frac{1}{8}$ audio jack to connect the microphone to the input of your preamplifier. You also can make a simple

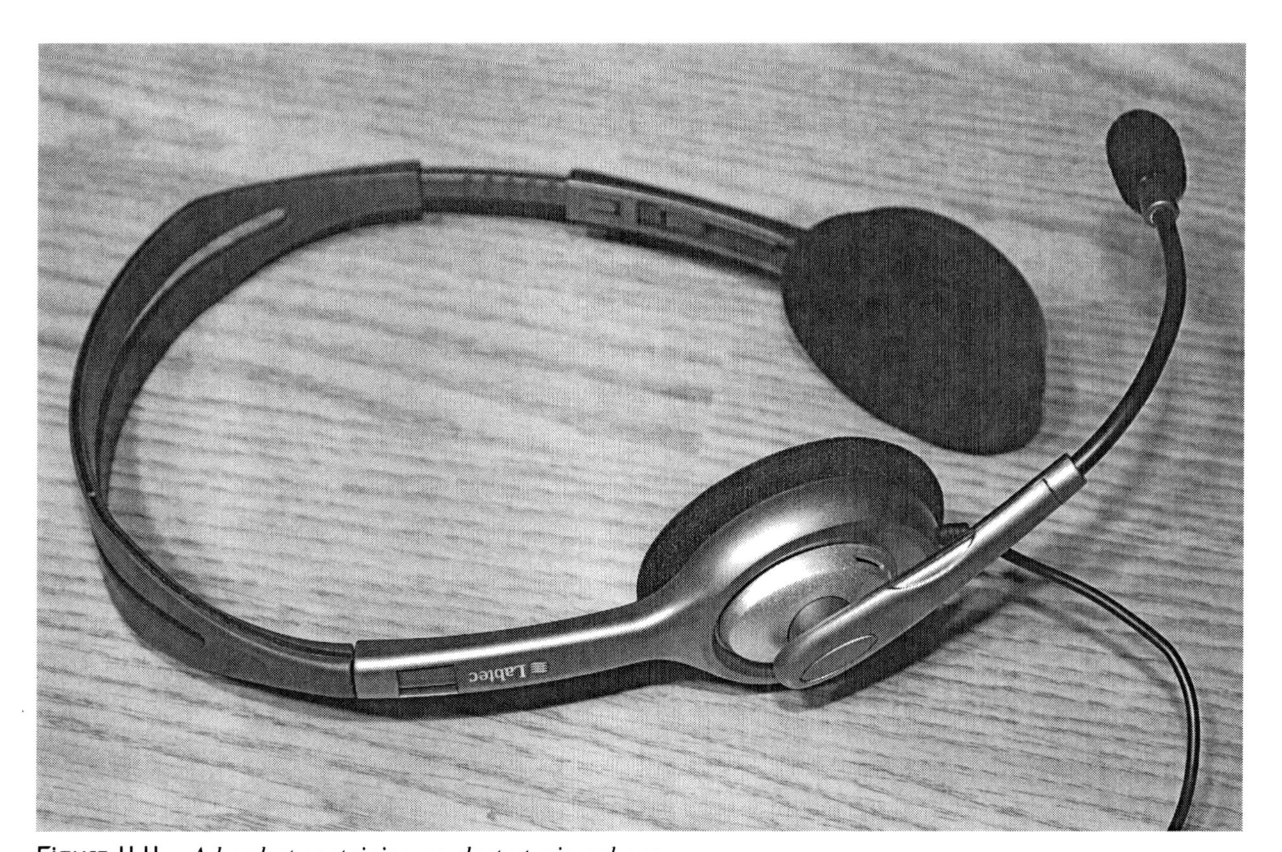

Figure 4-4 *A headset containing an electret microphone.*

headset by bending a coat hanger or simply use a bit of double-sided tape to hold the tiny electret microphone just under the subject's nose to get a good reading.

The breadboarded preamplifier circuit is shown in Figure 4-5 and is so simple that it can be built onto a very small circuit board or without any circuit board at all. The circuit will run on as low as 3 V and as much as 12 V, but a 9-V battery seemed most convenient and would power the preamplifier for a very long time. This preamplifier also makes a decent high-gain microphone amplifier for voice recording, so it can be used for many other experiments needing a sensitive microphone and preamplifier. You also can reduce the circuit further by replacing the variable resistor (VR1) with a fixed resistor with a value between 1 MW and 500K. The 1-MW resistor will give the amplifier the most gain possible, but it may be too much if you do plan to use the device to record voice or audio.

The preamplifier circuit had so few components that I decided to build it without a circuit board to save space and make the compact unit shown in Figure 4-6. I replaced the variable resistor with a 1-MW fixed resistor so that the amplifier would have full gain at all times. Because the LM358 IC was the largest part, it was used as a base to hold the other components into a "blob" circuit that was about half the size of a penny. I also could power the amplifier from a 3-V button cell and place the entire setup inside a bottletop so that only the output leads needed to come from the device. The completed preamplifier performed so well that I even used it for telecommunications over the Internet because it had better gain and clarity than the microphone connected to my laptop. Now all you have to do is find a way to log and analyze the data once the microphone is ready to record your subject's respiration.

The burst of audio data can be seen clearly in Figure 4-7 after recording a few hours of breathing using audio recording software. I use an older version of Sony Sound Forge, but just about any software-based audio recording software that allows you to see waveform data and time will do the job. The larger bursts of data

Figure 4-5 *Breadboarded respiratory monitor circuit.*

Figure 4-6 *Building a circuit without a circuit board.*

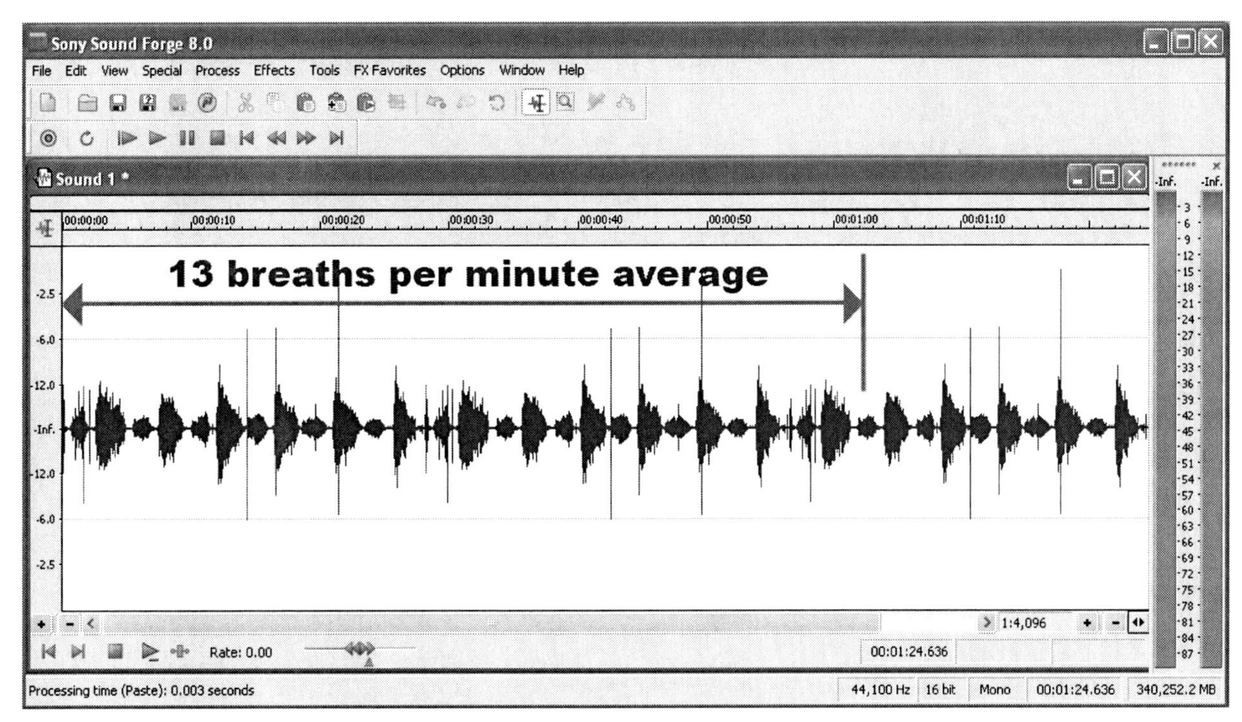

Figure 4-7 *Data bursts shown from the device.*

are exhalations, and the smaller ones are inhalations, so you can just count the bursts for 1 minute and then divide that value by two to get the average respiratory rate. If you are logging several hours through a sleep cycle, then you will easily see the changes to respiratory rate that occur during dream cycles. Breathing becomes shallow and erratic while we dream, so adding respiratory data to your dream laboratory certainly will help you to log your subject's cycles and trigger any external devices you are using. The audio-recording software is the most basic method of using the preamplifier to monitor respiration, and many other programs are available that can perform much more advanced analysis of these simple data.

You also could feed the audio output into an analog-to-digital converter built into a microcontroller and log the respiratory data directly as digital data. A comparator could be set up to read only the exhalation waveform because it is two or three times greater than the inhalation waveform. The circuit and source code presented in the light-sensing lucid-dream mask project in Section Two also could be modified easily to take its input from this circuit if you want to monitor data without the need of a computer. Add a few LED displays, and you could add a counter that simply counts breaths for as long as the circuit is running. Next, we will build a device to monitor heart rate.

Project Five

Heart Rate Monitor

Of all the body responses one could monitor, heart rate is one of the most important because it fluctuates greatly depending on our state of mind and physical condition. Heart rate during physical activity gives us a direct indication of our fitness level, and heart rate changes throughout the night can be clear indicators of when a dream cycle has begun. A heart rate monitor is also a very easy device to use for extended periods because it can be placed on the body in such a way that is not uncomfortable or in the way of any normal activity.

Our heart rate varies with age, gender, and physical condition, but the normally accepted range for adults is between 50 and about 100 beats per minute. While we rest, our heart rate is usually between 50 and 75 beats per minute, and it could climb to over 200 beats per minute while working the body to its near-maximum effort. Heart rate also becomes erratic during sleep cycles, so it can be used to trigger some type of dream experiment or be used in conjunction with many of the other dream-research devices presented in this book to track the dream cycles more accurately throughout the night.

The heart rate monitor project presented here uses a sensitive light-detecting resistor to detect small changes in the light beaming through your finger as your blood pumps through the tiny arteries in the finger. This device is not the same as the heart rate monitors typically used in hospitals because those use skin probes that detect changes in electrical activity as the heart beats. This heart rate monitor actually "sees" your pulse through the body, so it can be used on a finger, toe, or even your earlobe.

This project is a bit more involved than some of the other devices presented in this book, but it still can be built on a small breadboard with a few inexpensive components in a day or two. The completed heart rate monitor is very stable and as accurate as any you would find built into high-quality exercise equipment. Your heart rate will be displayed as a three-digit number on the LED readout, which updates its average once every few seconds. There is much room for improvement and modification owing to the simplicity of the microcontroller code, so you can easily adapt this device to just about any hardware or data-logging device.

The heart rate monitor is a combination of analog and digital circuitry, as shown in Figure 5-1. The LM324 quad op amp (IC1) forms a sensitive amplifier and a low-pass filter that will "lock" onto tiny variations in voltage that fall within the typical heart rate frequencies. The varying voltage comes from the light-dependent resistor (LDR), which changes a very small

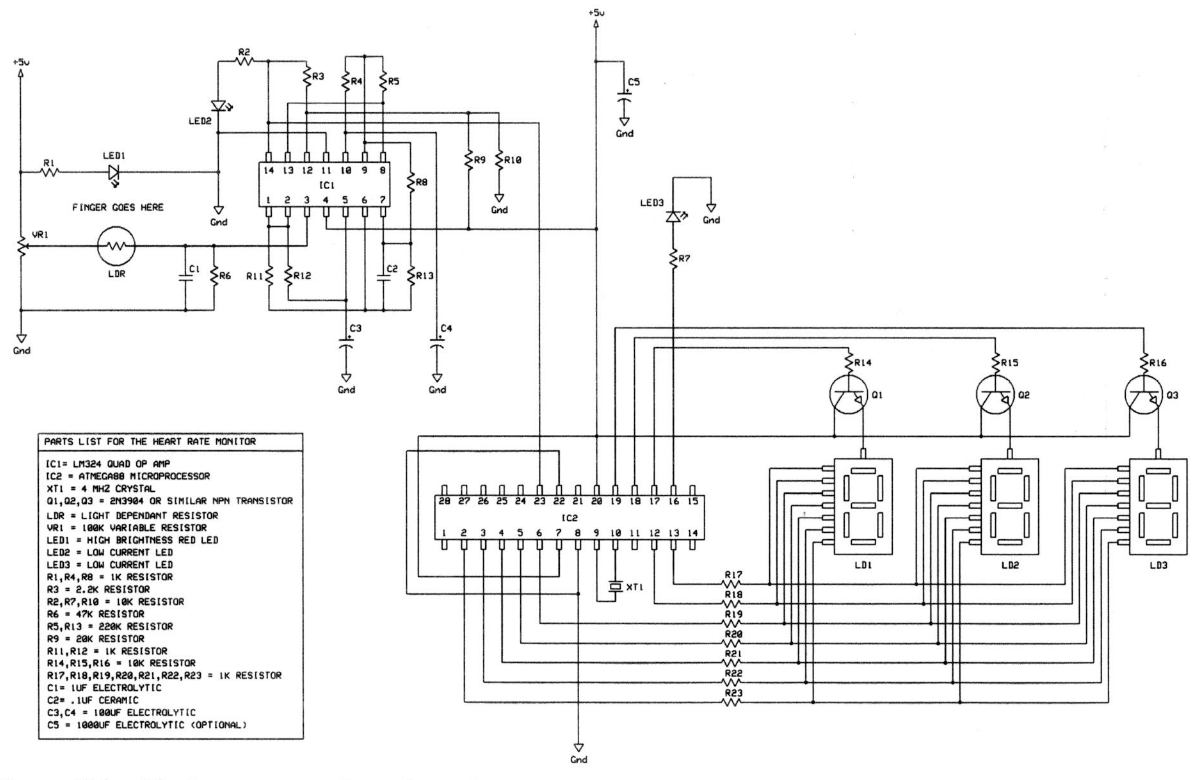

Figure 5-1 *The heart rate monitor schematic.*

amount each time blood pumps through the tiny arteries in between the visible red LED and the LDR surface. After the signal is conditioned by the low-pass filter, it is fed out of the op amp into the analog input of an Atmega88 microcontroller, where the program counts the beats and keeps a running average of beats per minute. The resulting heart rate is displayed as a three-digit number on the triple seven-segment LED displays. Because the microcontroller was added to this device after it was designed, you could decide to leave out the digital part of this project and take the output directly from the op amp and feed it either to an LED that will blink with each heartbeat or to some other data logger or device of your own making. The output from the op amp is very close to being TTL (transistor, transistor logic) compliant, so it could be adapted easily to drive most digital equipment requiring a 5-V TTL signal input. I recommend that you build the analog part of this circuit first because it is the most complex part of the circuit. LED2 will blink each time a heartbeat is detected, so this makes debugging the system much easier. When you get to the digital part of the circuit, another debugging LED (LED3) will blink each time the microcontroller receives a pulse signal from the analog section of this project.

The light-dependent resistor (LDR) is a commonly used semiconductor that can be found in practically every device that needs to respond to some change in ambient light levels. Street lights, security lights, light meters, and even dollar-store night lights such as the one shown in Figure 5-2 will contain an LDR. You also can purchase an LDR from most electronics suppliers, but the dollar-store night light is more convenient, costs about the same as a bare LDR, and gives you a few other bits for your junk box, such as a triac and a few resistors. An LDR is easy to identify because it will be visible to the light it must sense and will look like a tiny button

Figure 5-2 *A good source for a light-dependent resistor (LDR).*

with a snakelike pattern on its surface. This can be seen clearly under the tiny plastic eye in the figure.

The snaking track along the surface of the LDR shown in Figure 5-3 is connected to the two pins that exit the device. As light strikes the surface of the LDR, the resistance drops greatly, acting like an analog light switch of sorts. Notice how

Figure 5-3 *The LDR is easy to identify and extract.*

amazingly simply the circuit ripped from the night light really is—just an LDR feeding a triac through a resistor to turn on the 120-V ac light bulb. To remove the LDR, either use a soldering iron to release it from the circuit, or just bend the leads back and forth until the device breaks free.

The breadboarded heart rate monitor is shown in Figure 5-4, and as you can see, most of the real estate has been taken up by the triple-digit LED display and all its supporting hardware. The analog pulse-detection circuitry at the top of the breadboard consists of the LM324 op amp and a few resistors and capacitors. The large 1000-μF capacitor (C5) shown under the 9-V battery is necessary only if you are using a dc poser adapter or find that the pulse-detection circuit falsely triggers when the LEDs change. The analog circuit is so sensitive that the change in voltage from a single LED coming on actually can trigger the circuit into an oscillation if your power supply is not rock solid. The huge capacitor acts as a power-supply buffer to help filter out any tiny fluctuations caused by the LED display

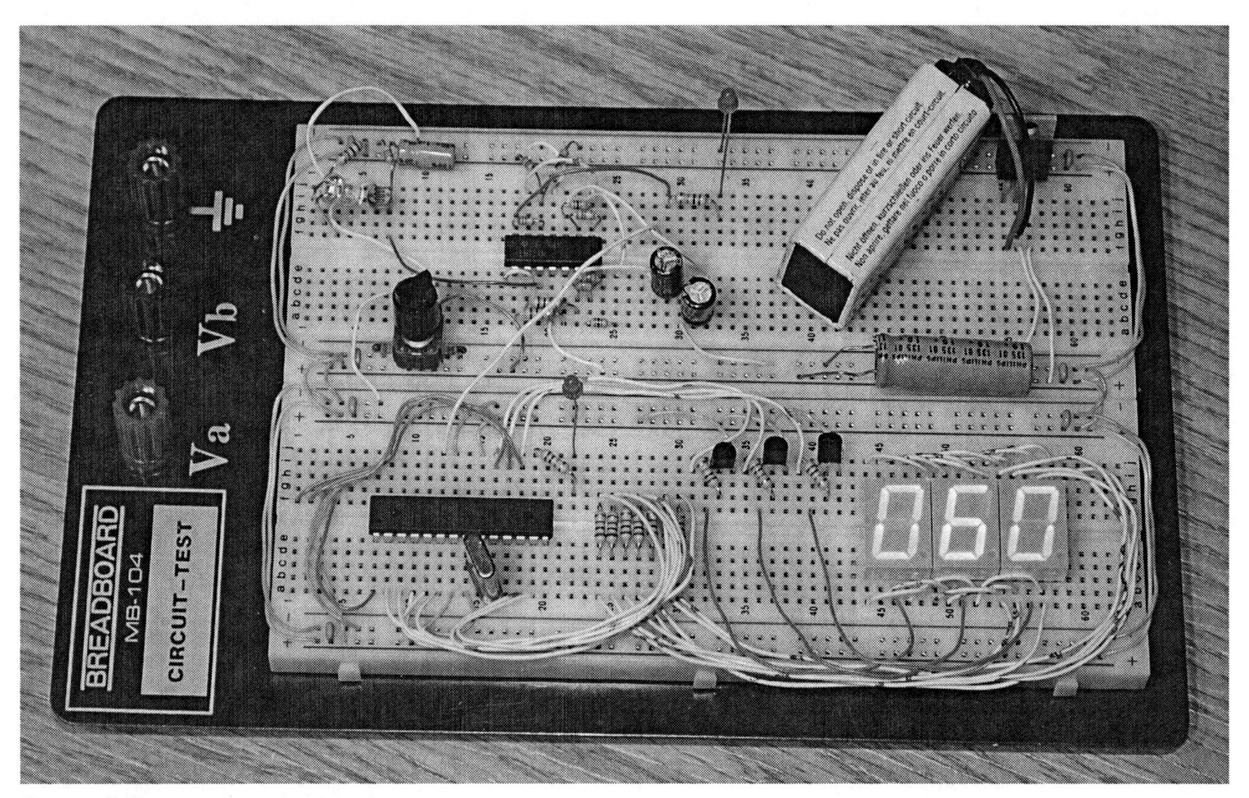

Figure 5-4 *The breadboarded heart rate monitor.*

circuitry. I found that the circuit worked fine without the large capacitor as long as it was running from a battery pack. Using a dc adapter did require the large filter capacitor to be included.

The component to the right of the 9-V battery is an LM7805 regulator to bring the supply voltage down to 5 V, as needed by the microcontroller. If you are building only the analog part of this circuit, it actually can be run from a supply voltage as high as 12 V directly. When first powered up, the LED display will read "060" and not change until a pulse is detected. Every 10 seconds, the average is recalculated, and then the readout is ramped up or down by a value of one on each following pulse. This slow ramp helps to filter out glitches in the average owing to movements of the body that might be detected as pulses. There is a lot of room to improve the microcontroller code because it was made to be short and simple to keep it printable.

To test the analog circuit, place a high-brightness red LED over the LDR, as shown in Figure 5-5, so that there is just enough room to place your finger in between the light and the LDR. You will have to keep your hand extremely still and at the same time take your own pulse using your other hand on your neck. When the

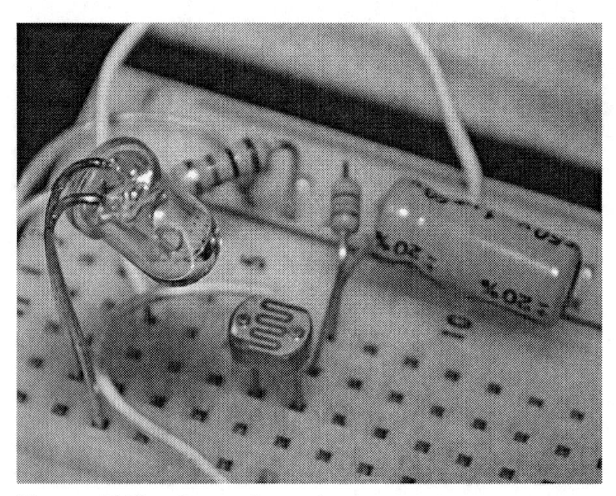

Figure 5-5 *Insert finger here.*

analog circuit is functioning properly, the pulse indicator LED (LED2) will flash almost at the same time that you feel the pulse in your neck.

If you just can't hold your hand still enough over the breadboard, move ahead and build the finger clip, which will eliminate false readings from accidental movement. You also will have to adjust the variable resistor (VR1) to some setting near its center of rotation to get the strongest reading on the pulse-indicator LED. The variable resistor biases the LDR in either the positive or negative direction, so you do not have to be overly concerned about the type of visible red LED you choose. The LED does have to be fairly bright and needs to be red because that is the color that passes through our skin the easiest. White light also will work, but green and blue LEDs will not work very well at all in this system.

I found that the system worked perfectly on the breadboard as long as I could keep my hand still enough not to send false readings to the analog circuit. If I rested the side of my hand on the desk and placed my thumb between the red LED and the LDR, then I rarely got a false reading. Figure 5-6 shows the result of 60 seconds of testing, getting a heart rate of 72 beats per minute, which jives perfectly with what I counted while taking the pulse on my neck. At this point, the analog circuit is fairly quirky when it comes to hand movement or even ambient light changes in the room, so some type of finger clasp will be needed to make the system much more stable.

The hair clip shown in Figure 5-7 just happened to fit around my finger or thumb perfectly, and it was not so tight that it became uncomfortable. There are hundreds of devices you could invent to attach the visible red LED

Figure 5-6 *Getting a new reading on the LED display.*

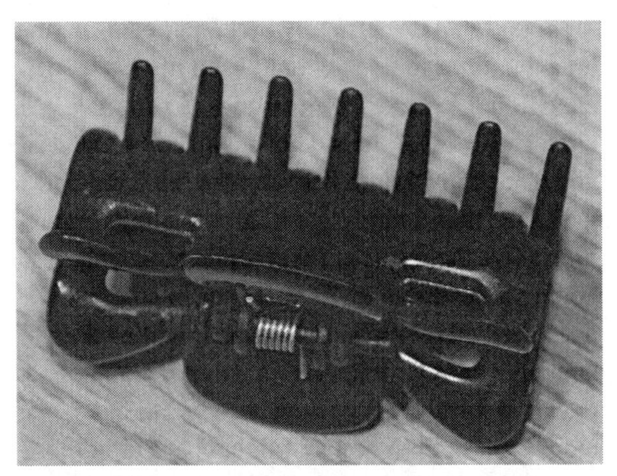

Figure 5-7 *This hair clip fits a finger perfectly.*

and LDR to your body, and remember that this device also will work through your earlobe. I intended to use my heart rate monitor for dream research, so it made the most sense to place the device on my thumb or toe so that it would not be annoying during the night. The method used to position the sensor should be both comfortable and secure enough not to move around easily.

To make the finger clip as compact as possible, the two ends of the hair clip are trimmed as shown in Figure 5-8. The clip now will work on any finger and on a large toe, although it seemed to work best on my thumb. I have poor circulation in my hands and feet, but since there is a large artery in our thumbs, the pulse was strong there no matter how cold my hands felt. Shooting the beam through the area of your fingernail also seemed to be the most effective place to get a strong and solid reading on the pulse-indicator LED.

The LDR and the visible red LED are placed at each end of the hair clip, as shown in Figure 5-9, so that the beam is directed onto the surface of the LDR once the clip is placed around a finger or thumb. My LDR had a small plastic bubble over the top of the detector surface, but some do not. I am not sure if moisture from your skin would affect an LDR without a lens, but if it does, then a small piece of tape or clear plastic placed over the surface of the LDR solves the problem easily. Don't worry if the LED beam

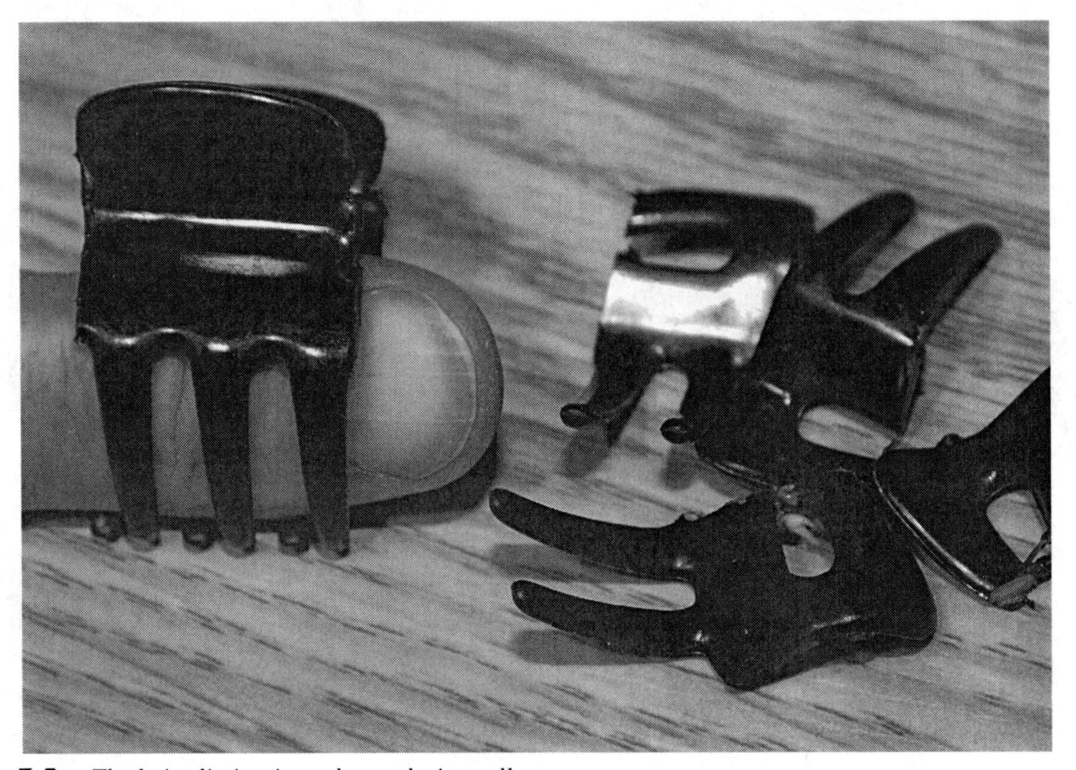

Figure 5-8 *The hair clip is trimmed to make it smaller.*

Figure 5-9 *Open jaws reveal the LED and the LDR.*

does not always line up perfectly onto the surface of the LDR because your finger will diffuse the light anyhow before it reaches the sensor. Since the analog circuitry is so sensitive, it takes very little light to actually cause the pulse detector to operate.

To ensure that the finger clip was working as well as or better than the original breadboard sensor, I placed it into the circuit for a reading. The variable resistor (VR1) actually needed a bit of a tweaking because the finger clip was much better aligned and passed more light to the LDR. After 10 seconds, the heart rate display climbed from the default power-on value of 60 to 89 beats per minute as my test subject held still under my massive camera lights for the shot. The elevated pulse was due to straining in the awkward position under 2000 Ω of light for over a minute while I got the shot I wanted.

It's amazing how little it takes to make our pulse shoot up from the normal, "doing nothing" rate. Just getting up from the desk was enough to add 5 beats per minute to my heart rate, although it quickly went back down to just over 60 beats per minute if I was not exerting myself in any way. Now the task of moving the circuit from the breadboard to a more suitable home will begin.

You will need a perf board that is at least 4 × 4 in to contain all the components in this project. However, take your time with the wiring, and start with the analog part of the circuit, just as was done on the breadboard. There are a lot of wires!

To keep the circuit board as small as possible, I replaced the LED displays with the smallest units I could find. If you have some talent with a soldering iron and PCB-making equipment, you probably could shrink the heart rate monitor down to the size of a matchbox, but I prefer the "old school" perf board method of prototyping because it is very easy to modify and correct errors.

Figure 5-10 shows the completed perf board ready to find a new home inside a plastic electrical box found at a local hardware store. The variable resistor (VR1) was replaced with a

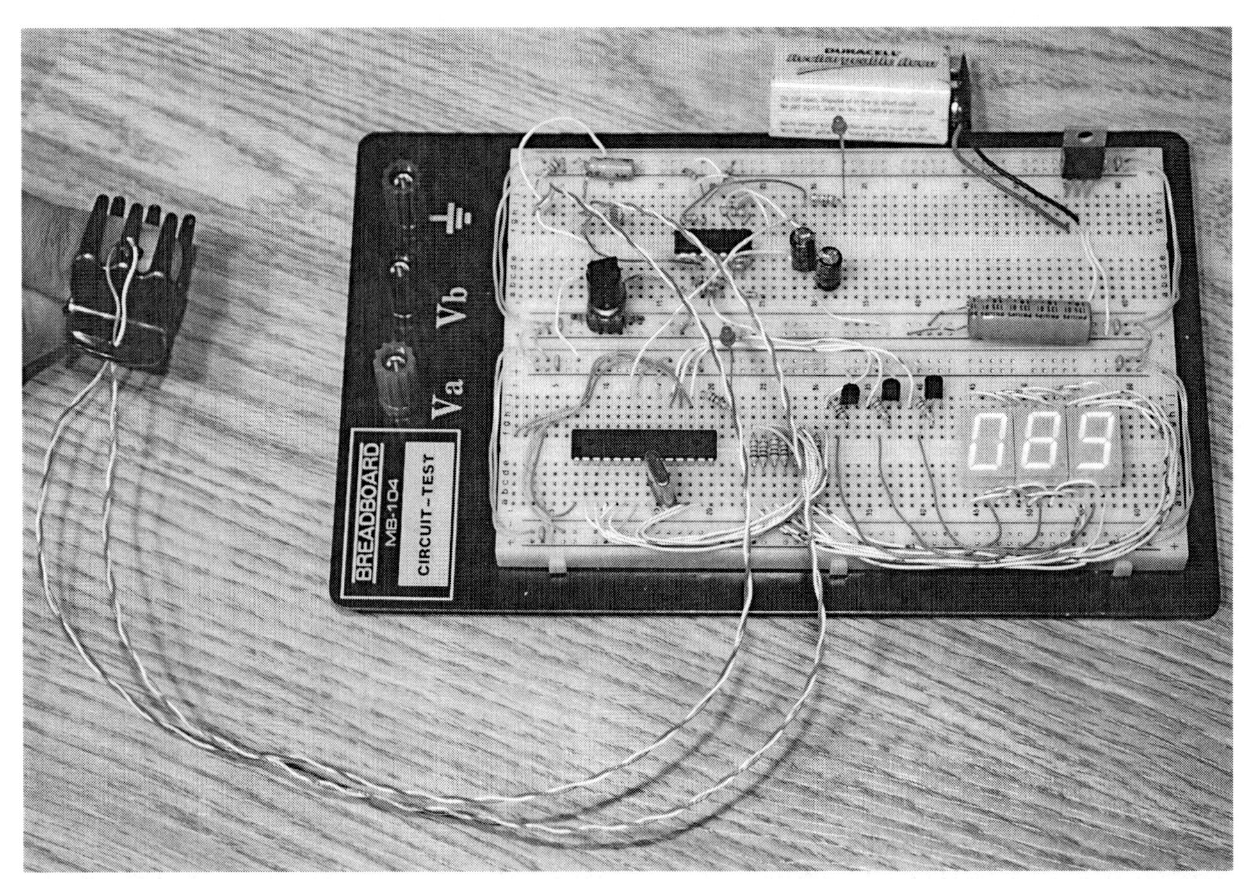

Figure 5-10 *Heart rate monitor on a perf board.*

board-mounted potentiometer because it only required adjustment once for the new clip-mounted optical sensor. If you plan to experiment with different sensor styles, then mounting the variable resistor externally might be a better option because it could require tweaking depending on the angle and amount of light hitting the LDR.

The completed heart rate monitor shown in Figure 5-11 performs as well as any store-bought unit but allows easy modification to adapt to just about any device you can imagine. There is plenty of room inside the cabinet for a small rf module, which would allow the heart pulses to be sent wirelessly to a computer for more advanced data logging. Because there are free IO pins on the microcontroller, it would be very easy to control some other device through a transistor or relay for further enhancements to the hardware. The source code is also very easy to adapt and expand because it was written to be as minimal as possible using the Basic language. Each block of code will be explained so that you can easily convert it to any microcontroller family or language.

Have a read through the complete Basic source code of Listing 5-1 provided in the appendix so that you can get an idea of how the program works. If you are an experienced programmer, then this trivial code probably is something you could write in 15 minutes from scratch, but if you have never written a program in your life, not to worry because Basic is called that for a reason and is easy to understand. The lines in the program listing that start with an apostrophe are comments, and I will explain the code in blocks after each comment. The compiler used is the Bascom AVR, and it is a very good compiler for the Atmega family and includes a very comprehensive instruction set.

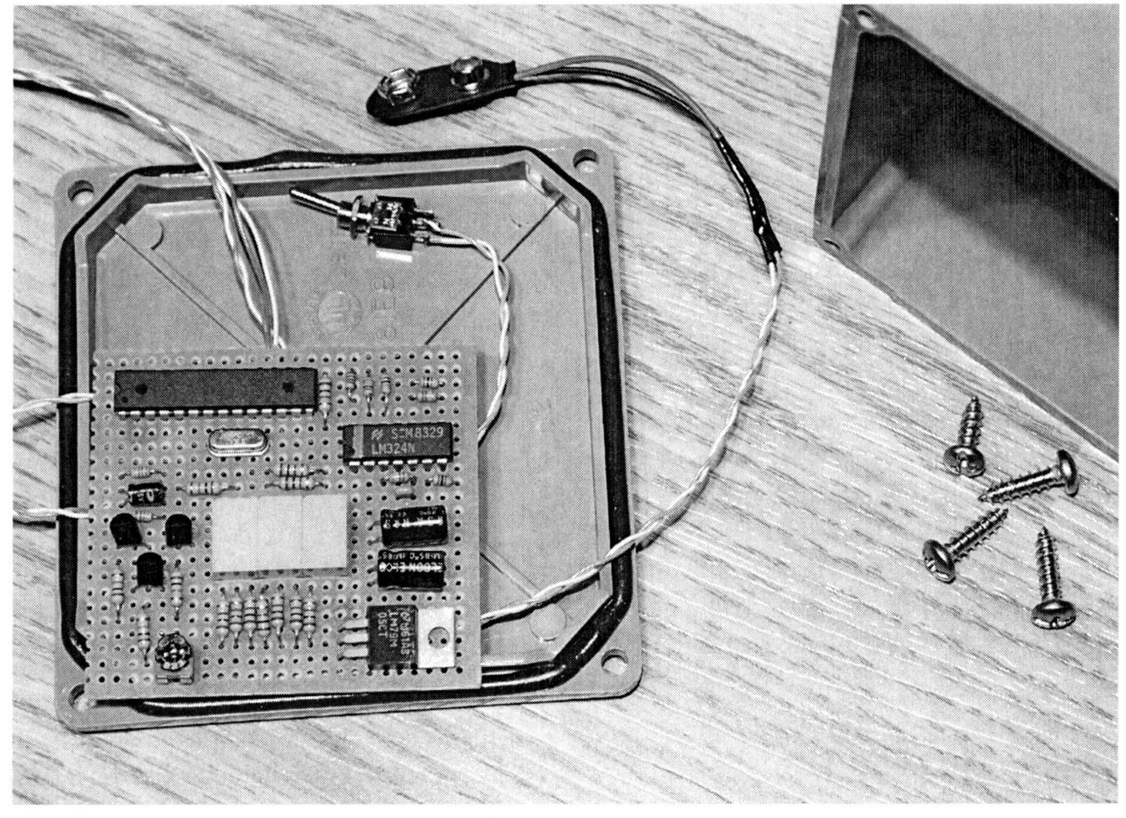

Figure 5-11 *Installing the components into a cabinet.*

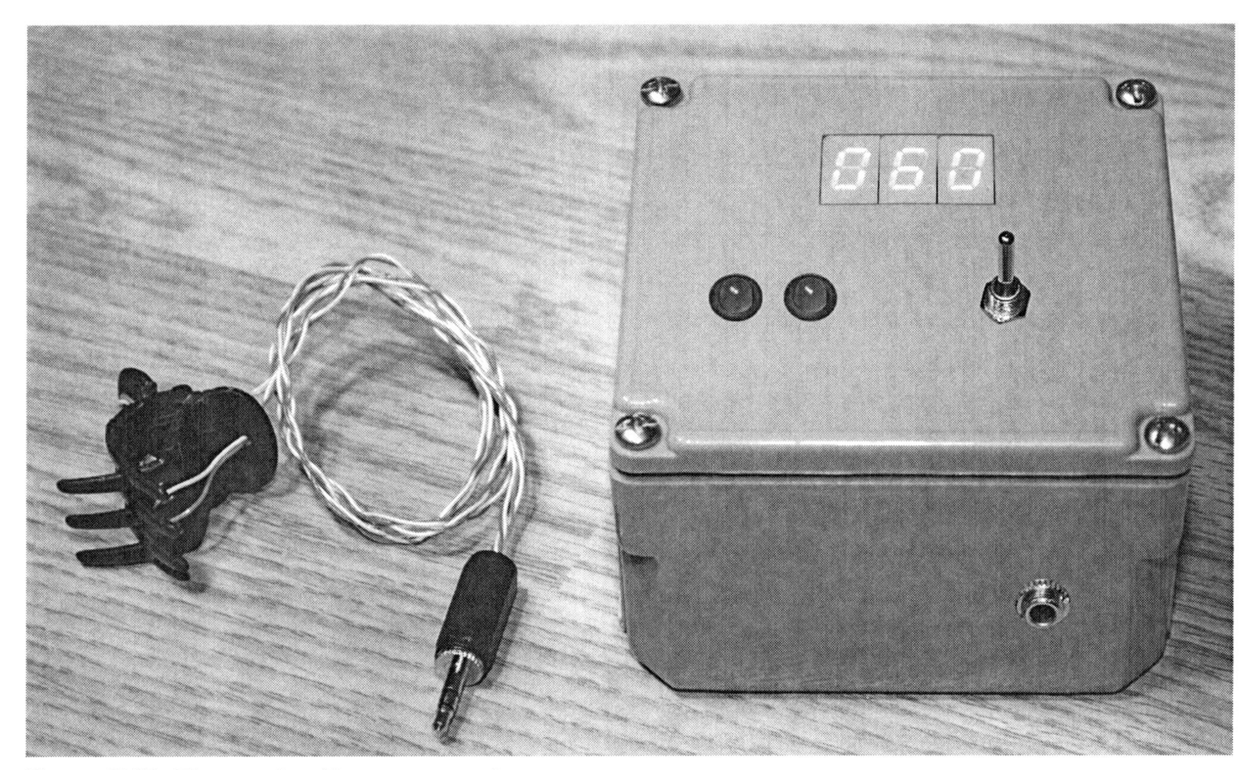

Figure 5-12 *The completed heart rate monitor.*

```
' DEFINE TARGET = MEGA88 @ 4MHZ
$regfile = "M88def.dat"
$crystal = 4000000
```

The code following this comment is required to tell Bascom that we are going to target the Atmega88 device and that our clock will be an external crystal resonator running at 4 MHz. Telling the compiler your clock speed becomes important when using commands that deal with timing-sensitive routines such as serial transmission or analog to digital readings. Defining the device also helps the compiler to generate user errors that have to do with IO pins. In this way, you can't accidentally try to toggle an IO pin that does not exist on the actual device.

```
' CONFIGURE IO PORTS
Config Portd.0 = Output
Config Portd.1 = Output
Config Portd.2 = Output
Config Portd.3 = Output
Config Portd.4 = Output
Config Portd.6 = Output
Config Portd.7 = Output
Config Portb.3 = Output
Config Portb.4 = Output
Config Portb.5 = Output
Config Portb.2 = Output
```

This block of code sets up the IO pins that will connect to the LED display outputs. You can change these around any way you like to make your wiring as simple as possible, but it is best to keep the LED digits to a single port for simplicity. Port D is being used here.

```
' DEFINE VARIABLES
Dim Led(10) As Byte
Dim A As Byte
Dim B As Byte
Dim C As Byte
Dim D As Word
Dim E As Word
Dim F As Integer
Dim G As Word
Dim H As Word
Dim J As Integer
Dim K As Integer
```

Basic uses *variables*, which are letters or words used to hold values. I like to use single letters such as *A*, *B*, and *C* for simple programs such as this one, but when you are working on a large, complex program, use of more descriptive variable names is recommended. "TIMER2" or "REDLED1," for example, would be descriptive variable names that make a lot more sense in a huge block of code.

Variables are also defined as the number of bits they are to contain, so in our code, "A" and "B" are 8-bit bytes that can contain a value between 0 and 255. Variable "F" is an integer that can range in value from –32768 to +32767. Variable "D" is a "Word" that can contain a value between 0 and 65535. Although you could just define all variables using larger data types, this is a waste of memory space and will slow down your code.

```
' DEFINE LED DIGITS
Led(1) = 8
Led(2) = 187
Led(3) = 97
Led(4) = 49
Led(5) = 178
Led(6) = 52
Led(7) = 4
Led(8) = 185
Led(9) = 0
Led(10) = 48
```

This block of code sets up an array of 10 values for the variable "LED." Although this may seem confusing at first, the values correspond to which of the seven segments will be lighted to display a particular number. To make this a little more confusing, the value in parentheses is actually one higher than the represented numerical value, and to light a segment, we want a low bit, not a high bit, so the value of "Led(9) = 0" says that to display the decimal number 8, we want all bits to be off. This means that all segments light up, and an "8" will be displayed. It can be a bit of a chore computing these values by adding port bits, but once you have done it once for your segment, the hard work is completed. Your values most likely will be completely different from mine as you "map" out the pins and segments on your LED displays.

```
' SET DEFAULT HEART RATE
K = 60
```

This sets up the variable "K" with a default rate that is close to the typical resting adult heart rate. By setting the average to 60 beats per mnute rather than starting it at zero, the calculated value does not have to "ramp" up so far to reach its final value. This will make more sense once you see how the heart rate monitor operates.

```
' START ADC RUNNING
Start Adc
```

The "Start Adc" command tells the compiler to include the code necessary to set up and initiate the onboard analog-to-digital converter on the Atmega88. This will allow us to read in an analog voltage and convert it to a value in order to detect changes in voltage from the LDR sensor.

```
' ********** MAIN LOOP **********
Main:
Do
```

Everything from here on is going to happen continually until the word "Loop" is reached, which causes program execution to start again where it first encountered the word "Do." This is called an *endless loop* because it never stops unless forced to by another command or an error.

```
' READ ADC VALUE
D = Getadc(0)
```

This command reads the analog-to-digital converter on pin 0 of the Atmega88 into variable "D," which is where the output from the LDR is connected. Since the ADC returns a 10-bit reading, values can range from 0 to 1024, which is why variable "D" needed to be a "Word," not a byte.

```
' GET ADC CHANGE SINCE LAST
F = D - E
F = Abs(f)
E = D
```

These three lines compare the current ADC reading "D" with the last known reading "E" by subtracting them into "F." The "Abs(F)" command changes the value in "F" to an absolute nonnegative value so that we only get the difference as a whole number, not a negative number, if "E" were greater than "D." Thus what this does is only give us the difference since the last ADC reading, not the actual value.

```
' HEART BEAT FILTER
If G > 0 Then G = G - 1
If F > 4 And G = 0 Then G = 40
```

This chunk of code acts like a crude low-pass filter that does not allow pulses to be detected that are much faster than what would be considered the upper limit for a human heart rate. Each time a new pulse is detected ("F > 4"), a counter is set to 40 and has to count down to zero before another pulse can be registered.

```
' HEART BEAT LED FLASHER
If G = 1 Then Portb.2 = 0
If G = 20 Then Portb.2 = 1
```

To aid in debugging the circuit and setting up the optical sensor, a pulse-indicator LED will flash on both the analog part of the circuit and the digital part of the circuit. This piece of code just looks to see if the heart beat filter has been reset and then turns an LED on and off for half the counter cycle.

```
' CALCULATE HEART RATE PER MINUTE
H = H + 1
If G = 1 Then
J = 9840 / H
H = 0
End If
```

To display the heart rate as beats per minute, a running counter "H" is incremented until the next pulse is detected. The value in "H" then is divided into a value (9840) that was calculated to give a fairly accurate result of beats per minute. This large number has to do with how many instruction cycles there are between the main loop of this program and took a bit of effort to get using an oscilloscope. Changing the auscultator frequency would completely throw off the calculated heart rate.

```
' SLOWLY ADJUST RUNNING AVERAGE
If G = 1 Then
If K > J Then K = K - 1
If K < J Then K = K + 1
End If<
```

To filter out the regain a little more, changes do not happen immediately but are "ramped" up or down on each pulse by a value of 1. By waiting until each pulse is detected ("G=1"), the final value "K" is compared with the calculated value "J" and incremented or decremented accordingly by 1. Without this ramping sequence, a single glitch or false detection would result in a wild and instantaneous change in the LED readout.

```
' DISPLAY DATA ON LEDS
If K > 99 Then
C = K Mod 10
B = K \ 10
B = B Mod 10
A = K \ 100
End If
If K < 100 Then
C = K Mod 10
B = K \ 10
A = 0
End If
If K < 10 Then
C = K
B = 0
A = 0
End If
Gosub Ledshow
```

Since the LED display contains three individual seven-segment displays and the actual value is contained in the "Word" variable "K," it must be broken down into three digits and sent to the display routine, which expects a value between 0 and 9 to be stored in each variable "A," "B," and "C." By using division and the "Mod" command, the values are taken from the variable "K" and broken into 1s, 10s, and 100s for the display routine.

```
' RESTART MAIN LOOP
Loop
End
```

The "Loop" command causes program execution to jump back to the main routine where the "Do" command was encountered. This is the end of the infinite loop, and all other commands beyond here must be called by either the "Goto" or "Gosub" command.

```
' ****** LED DISPLAY ROUTINE ******
Ledshow:
Portb.5 = 1
Portb.4 = 0
Portb.3 = 0
D = A + 1
Portd = Led(d)
Waitms 2
Portb.5 = 0
Portb.4 = 1
Portb.3 = 0
D = B + 1
Portd = Led(d)
Waitms 2
Portb.5 = 0
Portb.4 = 0
Portb.3 = 1
D = C + 1
Portd = Led(d)
Waitms 2
Return
```

This is the routine that displays a digit on each of the three LED displays. A little trick called *persistence of vision* is used here to switch between displays so fast that your eyes think that they are all on at the same time. As you can see, only one bit of the three "Portb" pins is on at a single time, and then the variables "A," "B," and "C" are sent to the display for only 2 ms each.

Because 2 ms is so fast, it appears that each display is on all the time, and the seven-segment lines can be shared, saving valuable IO overhead. Since Bascom array variables start at 1 not 0, the line "D = A + 1" adds 1 to the array pointer so that the value in "Led(d)" is the same as the decimal value we want. This conversion just makes it easier to understand the code, especially when trying to compute the segment bits from scratch the first time. Once all three displays have been lighted for 2 ms each, this routine just "returns" to where it was originally called.

I'm sure that you will find many ways to improve the program code and modify this project to suit your needs. With all those unused IO pins and code space, it would be easy to create a lucid dream-detection system by adding an 8-hour timer that looks for dream stages by comparing time with fluctuating heart rate. The addition of a relay or driver transistor could trigger just about any device, creating a complete dream-detection or -induction system that would work just as well as the eyelid movement–sensing device presented earlier. The ability to write Basic code and work with microcontrollers makes rapid prototyping a snap, so there really is no idea you can't set in motion with a little imagination and work. Enjoy!

For the last project in this section, we will continue to explore the reactions of the human body with an effective, sensitive lie-detector device.

Project Six

Lie Detector

A real lie detector, or *polygraph*, is a complex and sensitive piece of electronic equipment that logs several physical responses at the same time as the subject undergoes a battery of questions. A real polygraph, such as the ones a government agency might use, would usually include circuitry to monitor blood pressure, pulse, respiration, breathing rhythms, body temperature and skin conductivity, and even brain waves. The simple version of the lie detector presented here, based on the very old designs, measures only skin conductivity, although it could be used in conjunction with many of the other projects in this chapter to create a much more elaborate lie detector.

The lie-detector schematic shown in Figure 6-1 is surprisingly simple considering that it does a very good job of measuring the subject's skin conductivity. The two transistors form a very-high-gain amplifier that measures a tiny amount of current that will pass along the surface of the subject's skin as he or she holds onto the test probes. Since skin does not conduct if it is completely dry, the more the subject perspires, the higher will be the reading on the analog meter. Since it is a known fact that we perspire a little more when we lie, the tester can grill the subject and then look for small changes on the meter. Since the amplifier is so sensitive and we all have differing levels of skin conductivity, the variable resistor (VR1) can adjust the initial setting so that the meter is pointing to the middle of its range before the test begins. If the subject tries to relax too much, this will cause the meter reading to drop and could be interpreted as the subject trying to trick the test. A higher reading following a question would indicate that the subject may be lying and perspiring more than normal.

The circuit is so simple that you probably can forego the breadboarding process and build it right onto a bit of perf board. As shown in Figure 6-2, there are only two transistors and two resistors on the board. The only change you might need to make is the addition of a resistor in series with your analog meter if it happens to shoot all the way to the end even when the adjustment VR1 is all the way down. If your analog meter fails to move much at all, then you might have to increase the voltage by adding another 9-V battery in series with the current battery to make 18 V. Likely, your meter will be fine with this circuit, but it really depends on the impedance and rated voltage of the tiny coil inside. Most analog meters will respond to a very tiny voltage, even if their readouts say something like "kiloamperes" or "megavolts"!

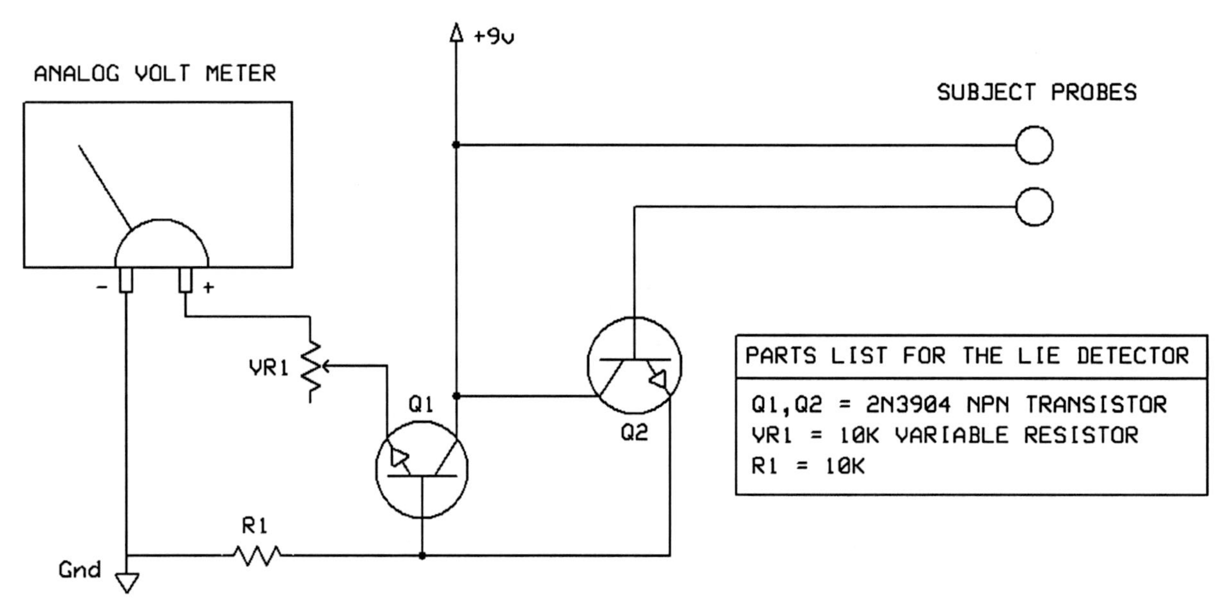

Figure 6-1 *The lie-detector schematic.*

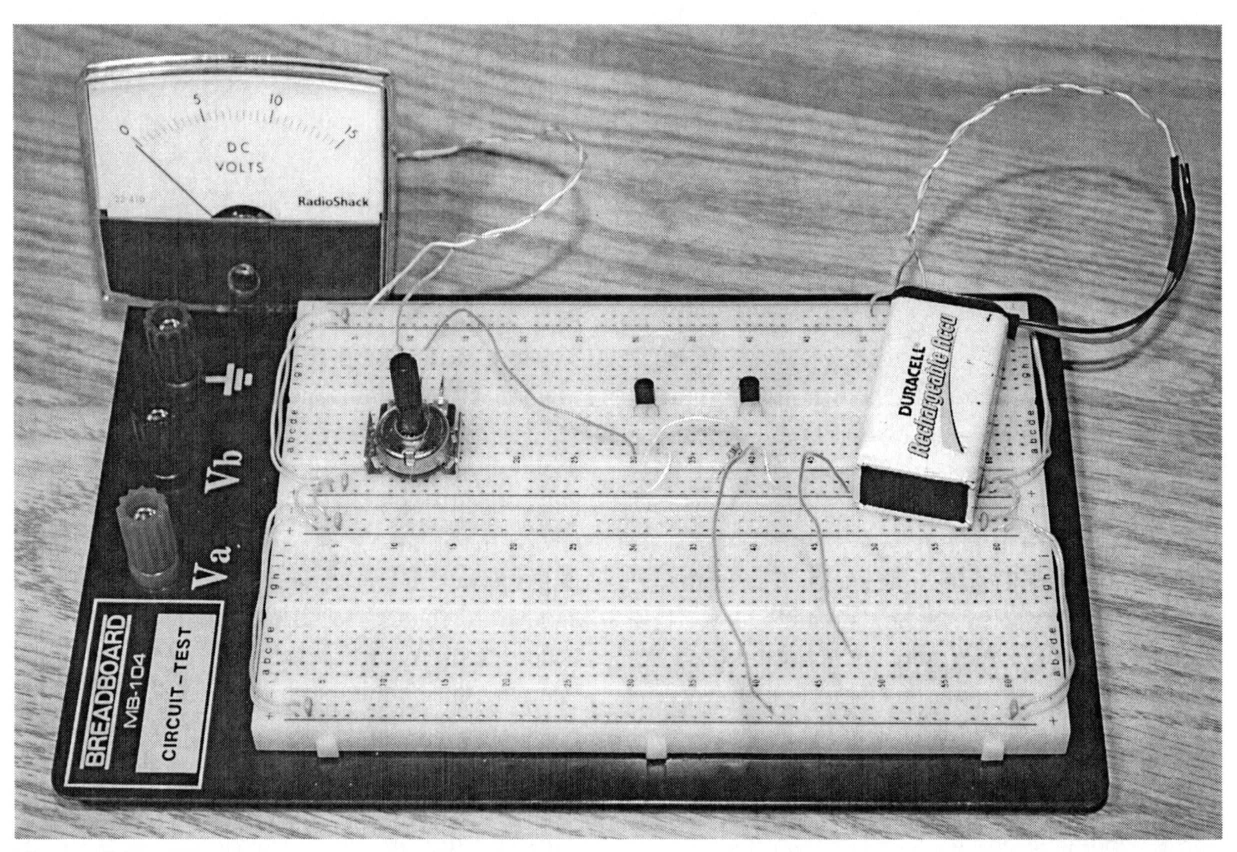

Figure 6-2 *The lie detector on a perf board.*

Analog meters can be quite costly to buy brand new, so you might want to do a little scrounging to find one to play around with. Second-hand shops often carry old stereo gear that may have one or more analog meters. Another source might be an old CB radio, battery tester, analog multimeter, or any other appliance from the 1970s or 1980s that had to display some value. Now that LEDs and semiconductors cost pennies, expensive mechanical display devices are very hard to find on mass-produced consumer electronics.

If you turn up the variable resistor and grab hold of the probe wires, your analog meter should jump to the end of its range in a hurry. If the meter fails to respond, try turning the variable resistor the other way in case you have the wiper lead connected in reverse. If there is still no response (unlikely), connect the probe wires together to drive the amplifier to its maximum value. If the meter still fails to move, then it is either not working or requires a lot more voltage to move the needle. If you are having the opposite problem and can't turn the variable resistor down enough to get the meter below the halfway mark, then add a resistor in series with the meter. A 1K resistor would be a good one to start with. Figure 6-3 shows the response you want to get—the meter at the halfway point after adjusting the variable resistor. You should be able to make the meter swing all the way over by wetting your finger or just by putting a lot of pressure on the probe wires.

Your analog meter likely will have a scale measuring amperage, voltage, or some other value unrelated to our biologic subject. It is actually very easy to make a replacement readout plate just by cutting out something you made in a computer paint program and then affixing it to the original plate. The plastic cover can be popped off by placing a small screwdriver or blade in the tiny slots to pry it off the backing plate, but be careful not to damage the sensitive meter or any of its moving parts once the cover has been removed. Figure 6-4 show the replacement readout plate I made that is a bit

Figure 6-3 *Testing the range of the analog meter.*

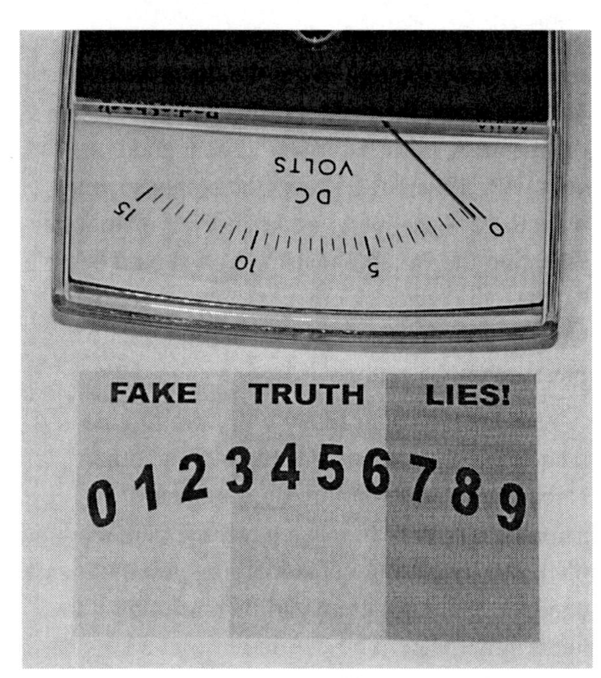

Figure 6-4 *Making a replacement readout plate.*

more on the "funny" side of things because I planned to use my lie detector just for kicks. You can make your readout say whatever you like, but try to create three distinct zones so that you can have your needle point to the center point when adjusted properly. Remember, a low reading will indicate either a bad connection between subject and probes or a subject trying to fake the result, and a high reading will indicate above-normal perspiration or a lie.

Once you have your clear plastic cover pried from the backing plate, simply glue or tape your new display over the old one, as shown in Figure 6-5. You do not have to remove the original reading plate because there will be plenty of room between the needle and the original plate to insert your paper display as long as it is glued down flat. I used a small dab from a glue stick to hold the new readout in place so that it can be removed easily at any time without damaging the original plate. Again, be careful when working inside the meter because that tiny needle is very easy to bend or break.

Although you can connect the probes just about anywhere on the subject's body, the old "sweaty

Figure 6-5 *Gluing the new display over the old one.*

palms of a liar" factor certainly applies to this device. I found a few steel rings at the local hardware store and then soldered wires to them, as shown in Figure 6-6, to make a good pickup for the subject's palms. The goal is to affix the probes to your subject in such a way that they make good contact with the skin but do not allow the subject to change the pressure on the probes, which would cause a false readout. By placing the rings on the subject's palms as he or she holds the hands palms up, you get a good connection without allowing the subject to grip or manipulate the pressure between the metal and the skin. If your subject has particularly dry hands, you can try placing the rings further up his or her arms or use elastic to apply a fixed amount of pressure to the probes.

For such a sparsely populated circuit board, there are sure a lot of wires coming from it. The analog meter, power switch, battery, probes, and variable resistor are all connected to the functional circuit shown in Figure 6-7, ready to be installed in some type of cabinet. I also decided to connect my probe leads through a ⅛ stereo jack so that I could create multiple probe sets to allow the use of variously shaped probes. You also could connect multiple probes to your subject as long as they are the same on each side of the body. On a real polygraph machine, many probes are connected to the subject for more accurate results.

There are two ways you can mount the analog meter in your project box: behind a square hole or over the top of the lid with only the "can" stuck through the lid. Making a square hole that looks good requires a bit of patience as well as the notching tool shown in Figure 6-8. To use a notching tool, trace the area to be cut out, and then drill a hole in each corner large enough to insert the tool's cutting blade. The notching tool allows you to "nibble" out small square bites from thin metal or plastic cabinets to create square or straight-edge holes. An easier way to mount an analog meter is just to cut a hole large

Figure 6-6 *Making probes for the subject's hands.*

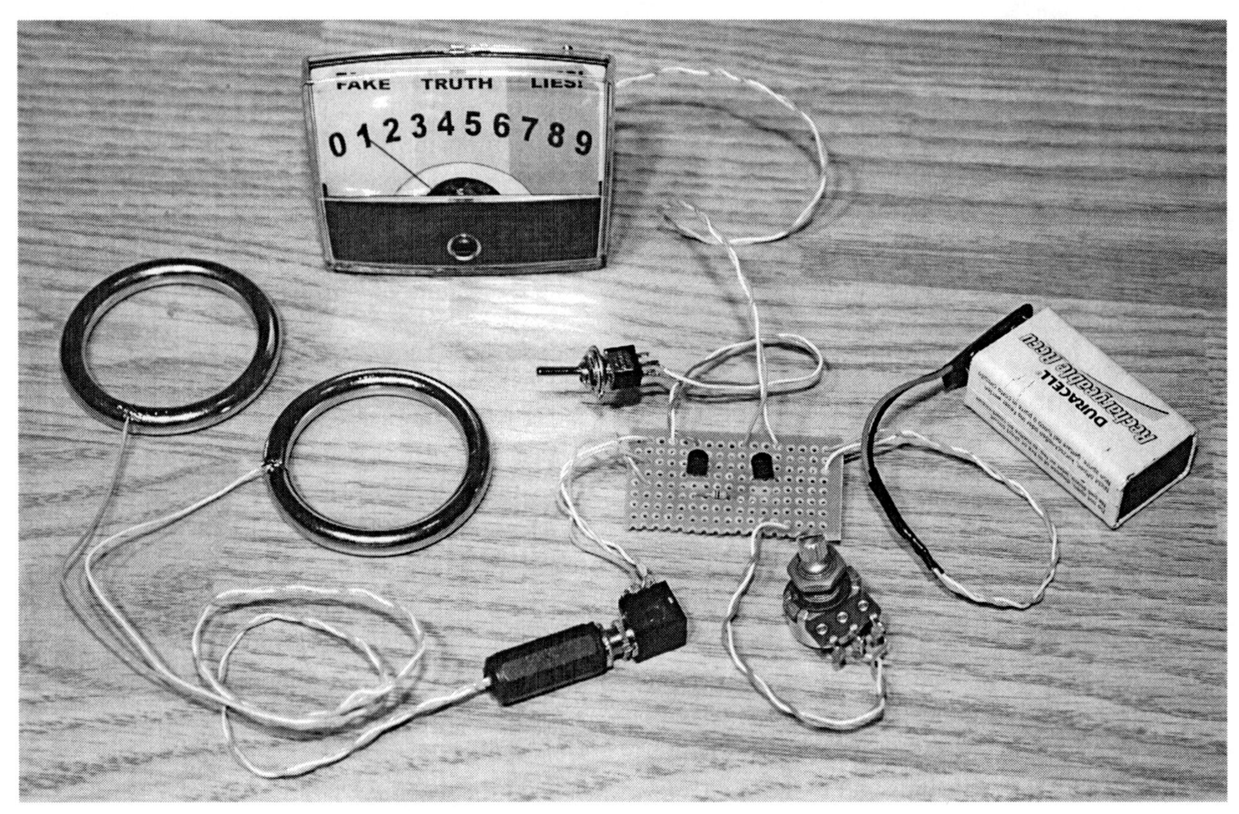

Figure 6-7 *The circuit ready to be installed into a cabinet.*

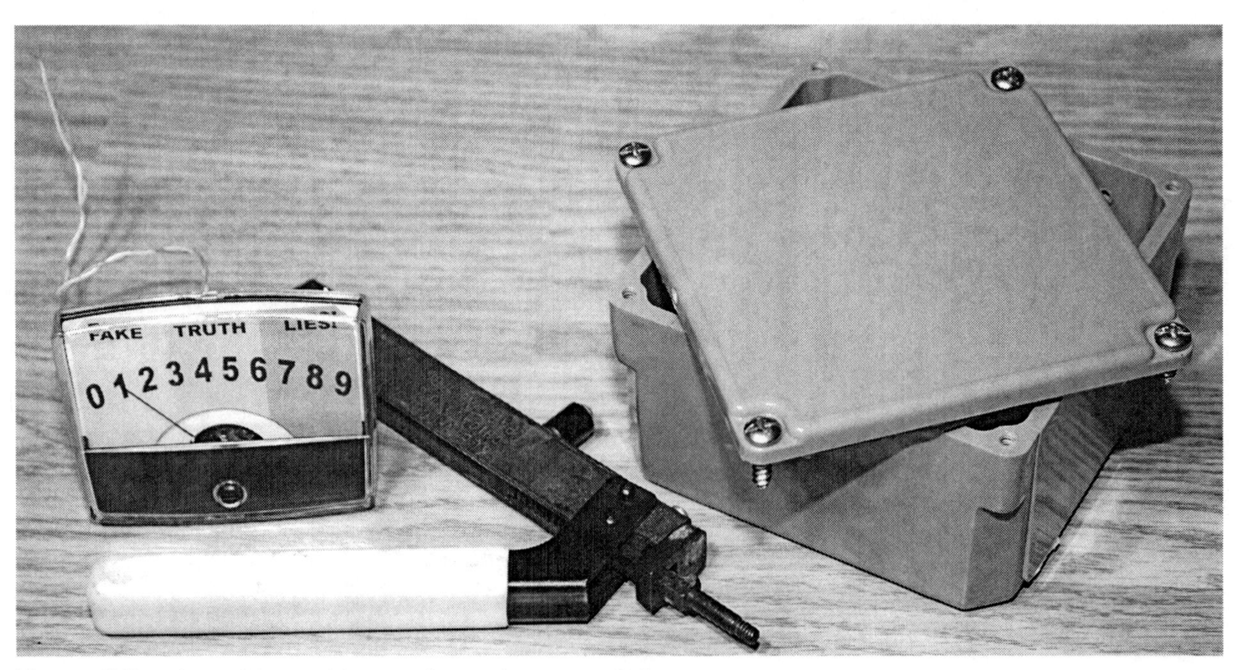

Figure 6-8 *A notching tool is great for cutting square holes.*

enough for the mechanical body part to fit through the lid, and let the display area cover the hole. This method is simple, but then you see the nondisplay area under the meter, which does not look as professional. Since the plastic cabinet I planned to use was much too thick for the notching tool, I chose the easy way out!

Like Agent Mulder from the *X-Files* once said, "The truth is out there." And with your new lie detector, you can cut through the deception and misinformation. After a bit of hole drilling, the completed lie detector shown in Figure 6-9 was ready for action. Of course, with my analog meter reading "Fake, Truth, or Lies," it is obviously more of a party gag than a tool I intend to use to interrogate my "enemy spies." You actually could combine this device with some of the other biofeedback devices shown earlier in this chapter to create a much more comprehensive lie-detection system, but you also will require the knowledge needed to decode all the feedback received from your subject. The art of using a polygraph is so controversial that some countries do not even consider it a valid test, and often those with knowledge of how the device works can learn to fool the tester.

I'm sure that you will find some useful application of your simple lie-detector unit and have some fun at the same time. If you make the unit look "real" enough, then it might be just as effective at "extracting" the truth as its big brother, the polygraph machine. Sometimes the risk of having a lie exposed is enough! In Section Two we will explore the dreaming world.

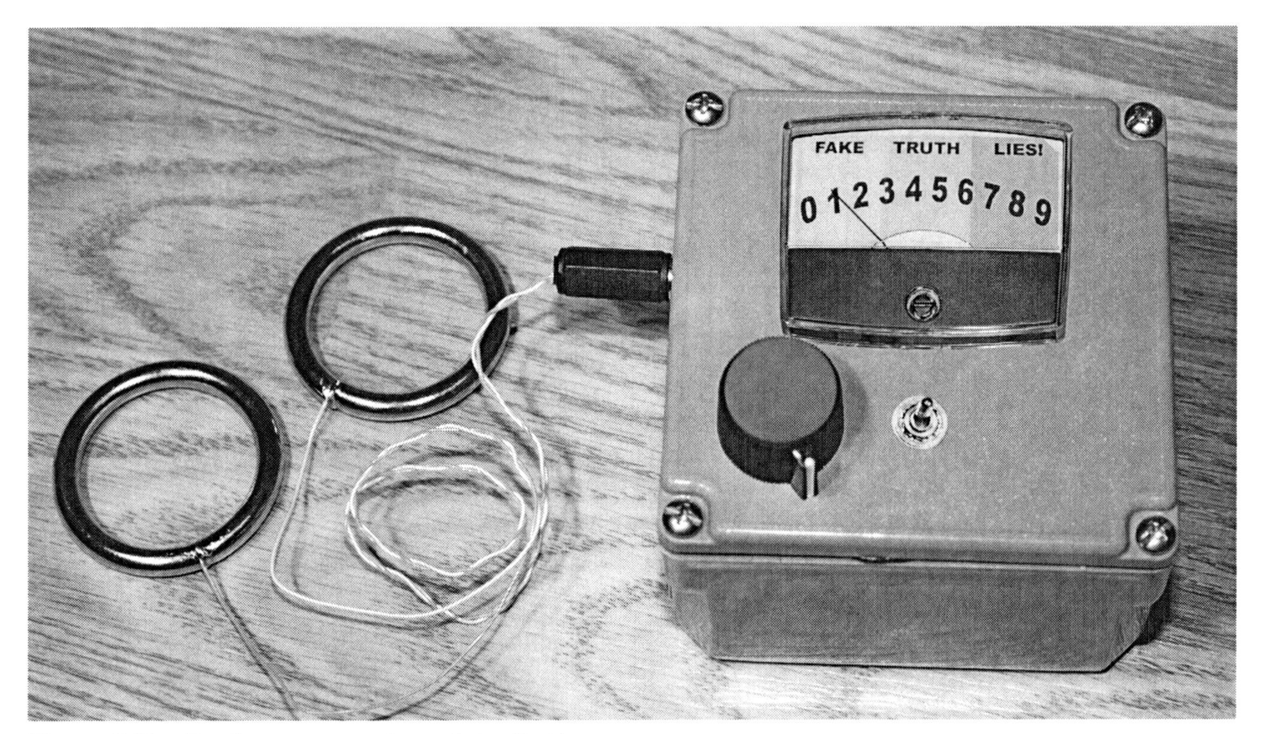

Figure 6-9 *Ready to uncover the truth (or lies)!*

Section Two

The Dreaming World

This section explores the other half of your life that unfolds when the lights are out and you are in that strange place called *dreamland*. You might think that a dream is something that we have no control over, but this is not the case, as will be shown with some of the devices presented here that will allow you to link your sleeping mind with your conscious mind. Imagine being able to enter a world where there are no limits or rules and where even the laws of physics are as elastic as the strange and wondrous atmosphere that surrounds you. Having conscious control over our dream world is something we are all capable of with a little training and help from the electronic world. So get ready to become an "oneironaut" and travel through your dream world, painting the scenery at will and allowing your wildest fantasies to unfurl because this section is all about connecting with the other half of your life.

Project Seven

White-Noise Generator

White noise can be best described as a hissing sound that an FM radio makes when tuned to a place on the dial where there is no station or static. White noise is a smooth hiss consisting of all audio frequencies played back at random simultaneously, which is why it has no discernible tone. White-noise generators have many uses, some of which include the testing of audio equipment, random-number generation in digital circuits, jamming audio bugs, and many medical uses that we will be able to test and experiment with using our simple device.

White noise is known for its calming effects, which is why many people find it easier to sleep with a fan running or some other source of random noise similar to white noise. Ocean waves, trickling water, and wind through trees are a few other examples of white noise, although they are not as steady as the white noise that will come from our home-built device. Because your ears cannot "lock" onto any specific frequencies or patterns in white noise, the device has the effect of shutting off your ears to distracting noises, just as being outside in the middle of a fresh snowfall on a bright day will give you snow blindness. Sensory deprivation using white noise even has been used as an interrogation aid to temporarily confuse subjects before questioning.

White noise is also used as a treatment for tinnitus and hyperacusis by distracting a patient's hearing so that it always hears something yet at the same time really hears nothing. Maybe you will want to try white noise as a sleep aid in your noisy apartment, or perhaps you want your next hacker gathering to be completely secure so you want to flood your windows with white noise to defeat the laser-bounce eavesdropper. Whatever your plans may be, this extremely easy-to-build white-noise generator will fit the bill and even can be made as a dual-channel stereo system.

The white-noise generator schematic shown in Figure 7-1 is so simple that you might think there is something missing, but the unit works very well and exploits the fact that transistors generate random noise when reversed biased. The reverse-biased emitter-to-base junction of Q1 is fed into a 1-Ω audio amplifier integrated circuit (LM386) so that it can power a speaker or a set of headphones. Because many transistors will act differently in this application, I have added a 100K variable resistor (VR1) to allow for some "tweaking" of the white noise for best results. You can use practically any small-signal NPN transistor in this circuit, and I have found the commonly available 2N3904 to give great results. Another variable resistor (VR2) controls the level

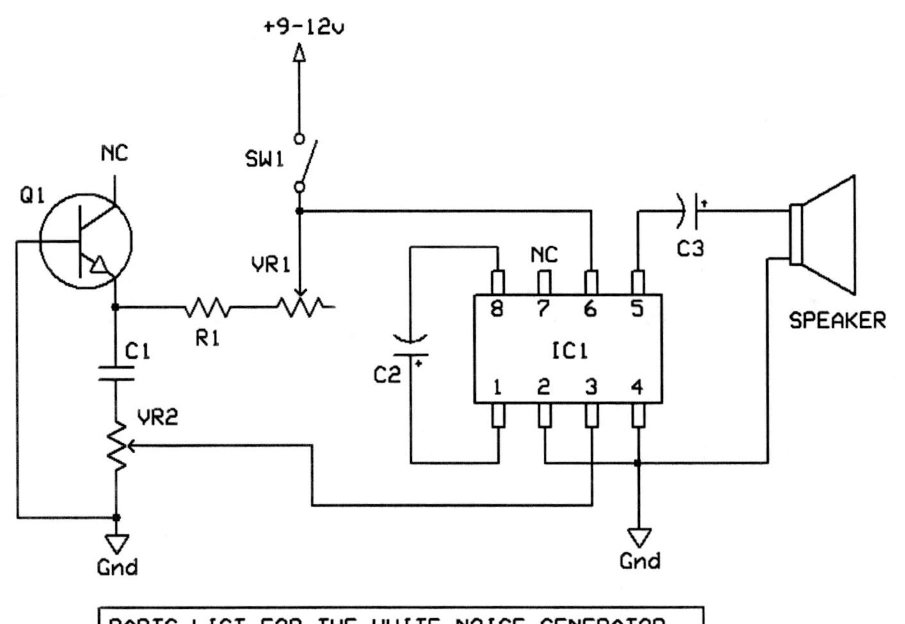

PARTS LIST FOR THE WHITE NOISE GENERATOR
IC1 = LM386 AUDIO AMPLIFIER
Q1 = 2N3904 OR SIMILAR NPN TRANSISTOR
R1 = 10K
C1 = .1UF (104)
C2 = 10UF
C3 = 220UF
VR1 = 100K VARIABLE RESISTOR
VR2 = 10K VARIABLE RESISTOR
SW1= SPST TOGGLE SWITCH

Figure 7-1 *White-noise generator schematic.*

of noise reaching the audio amplifier so that you can adjust the level to suit your needs. Even though the LM386 audio amplifier is just a small 8-pin intergrated circuit (IC), it actually can drive any size speaker you like and is probably more than loud enough to fill a room full of clean white noise.

For a very relaxing white-noise generator that will seem to send noise in all directions, you can build the stereo version shown in Figure 7-2 simply by doubling up the original version and sending the outputs to a pair of stereo headphones, speakers, or the line inputs on your audio amplifier. Because there are two independent noise generators running, the sound will seem to come from all directions, which works amazingly well for audio-deprivation experiments where the subject is wearing headphones or is placed directly between the two speakers.

If you do plan to build the stereo version, try to keep all parts the same make and value so that the sound is even on both channels. Since some transistors may behave slightly differently in this circuit, they should both be exactly the same. To simplify the stereo circuit a bit further, you could replace the LM386 with a stereo amplifier IC to reduce the parts count a bit. If you plan on feeding the output directly to a line input on an amplifier, then you actually can eliminate the entire amplifier section and simply feed the output from the center tap of VR2 directly to your amplifier's input.

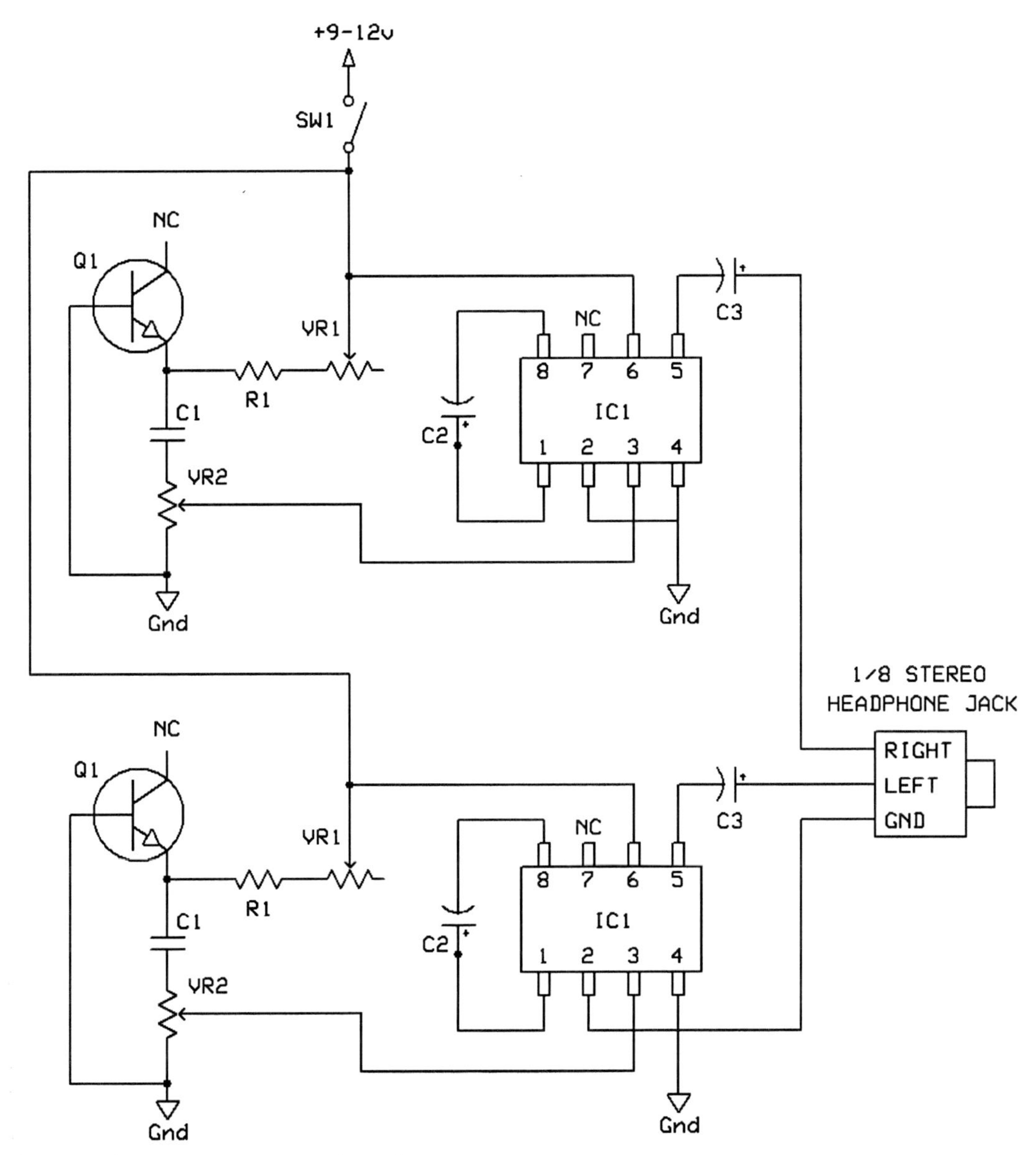

Figure 7-2 *Stereo white-noise generator schematic.*

Figure 7-3 shows what white noise looks like on an oscilloscope or computer input. Since white noise is comprised of every possible frequency at the same time, there is no discernible pattern or tone that can be detected, which is why your mind finds it so relaxing to listen to. White-noise generators like this are often used in digital machines that need to create some kind of chaos for a random output. Digital encryption and lottery machines are prime examples of using white noise as a source for random information because digital circuits cannot generate truly random sequences that never repeat.

The simple circuit can be built on a solderless breadboard (Figure 7-5) in a few minutes. As soon as you power up the circuit, a nice clean hiss should be heard from the speaker if everything is working well. If your noise sounds more like a crackling or spitting oscillator, then try moving VR1 to tweak the sound a little bit. If your noise is still not smooth and crackle-free, just drop in another transistor, and listen to what you get. The

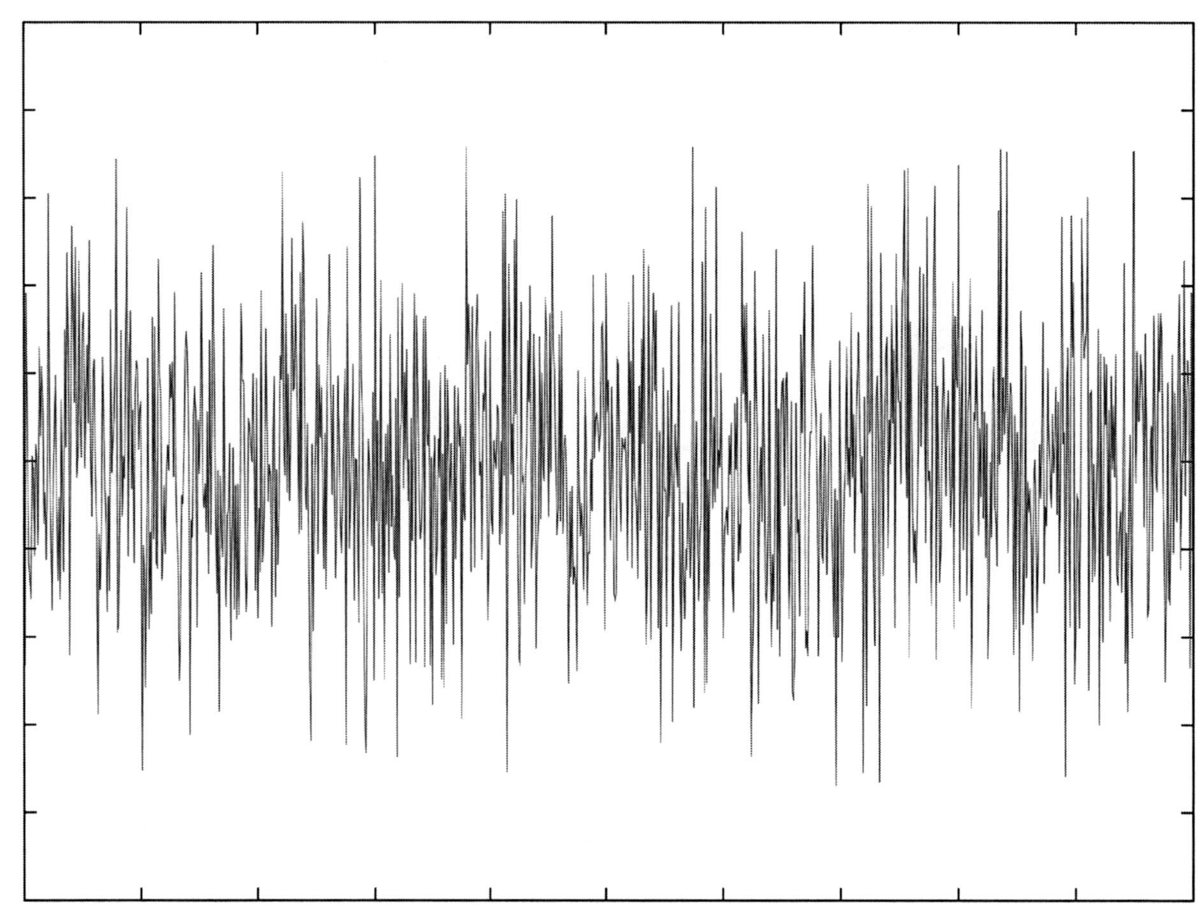

Figure 7-3 *White noise is a random wave.*

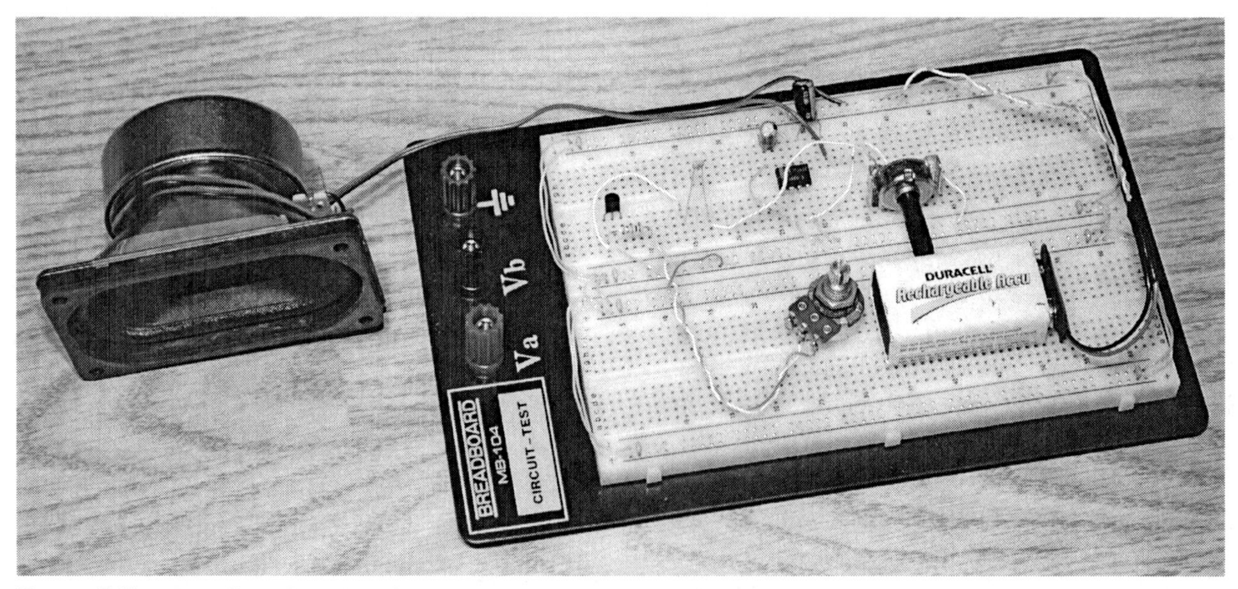

Figure 7-4 *Breadboarding the white-noise generator.*

optimal white noise is a very clean hiss much like a radio with no station or the sound of air leaking out of a tire. Since the LM386 audio amplifier is happy on any voltage from 9 to 18 V, you might want to try a 9-V battery for the absolute cleanest possible output. A dirty power supply will add hum into the circuit.

If you are planning to use headphones, turn down the volume control (VR2) at first because the output of the audio amp can be very loud. If you find the output to be much too loud, remove capacitor C2, and that will change the gain on the LM386 from 200 to only 20, making the output much quieter.

The perf-board version of the white-noise generator shown in Figure 7-5 is extremely small, so it can be installed into just about any cabinet. If you are happy with the "flavor" of white noise after adjusting VR1, then you simply can measure the impedance when it is set correctly and replace it with a fixed resistor to simplify the circuit even further. I decided to leave the tweaking adjustment because it worked somewhat like a single-band equalizer to adjust the sharpness of the white noise.

On a single 9-V battery, the white-noise generator will run all day, although you also could use a 9- or 12-V direct-currebt (dc) adapter for batteryless operation if needed. The 4-inch-diameter speaker shown in Figure 7-5 was fine for a small room, but for a larger room or a more bassy sound, you probably would want to use a larger speaker or one already mounted inside a cabinet. A 10-in speaker in a cabinet will make the white noise much richer and more like wind rather than a hissing radio.

The stereo version of the white-noise generator shown in Figure 7-6 is great for relaxing or messing around with auditory-deprivation experiments. This version has dual noise generators, and both audio amplifiers have capacitor C2 removed so that the output is not too

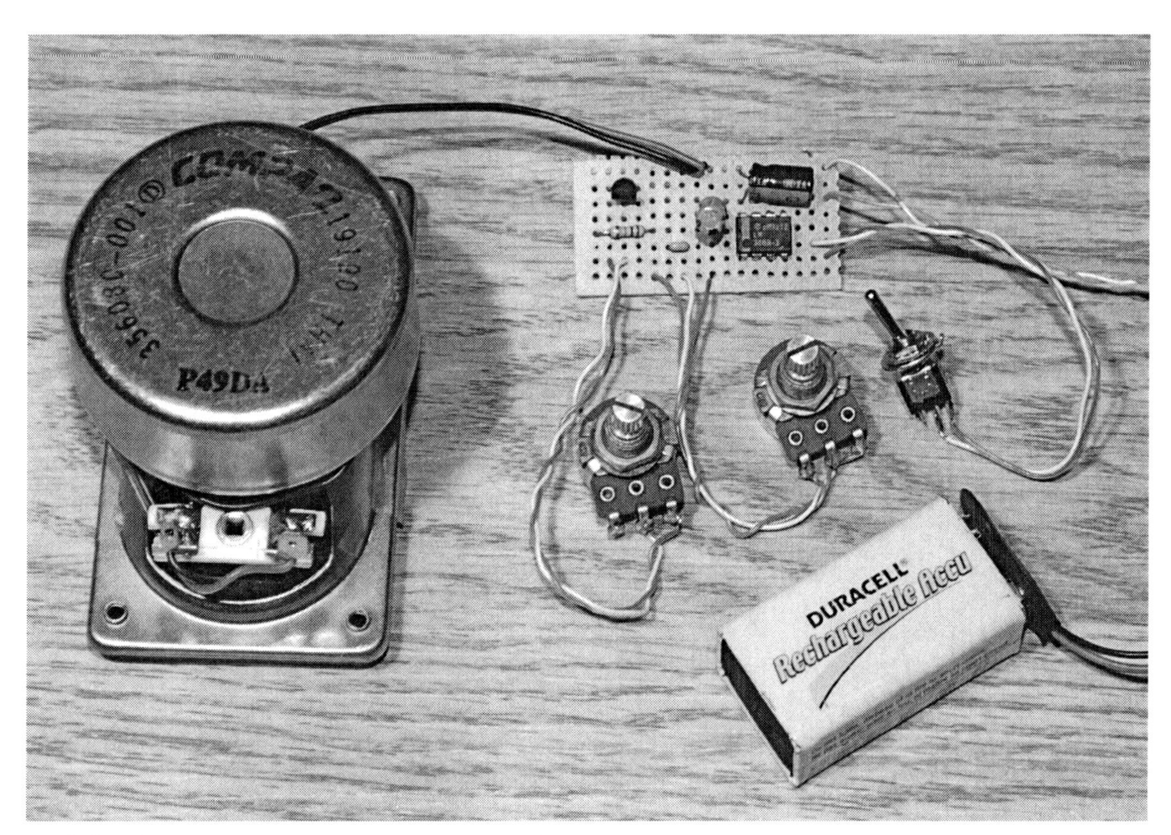

Figure 7-5 *White-noise generator on a perf board.*

Figure 7-6 *Head-mounted stereo version.*

Figure 7-7 *A larger version with more bass response.*

loud for the headphones. To connect this unit to a stereo amplifier for big-room sound, you just need a ⅛-inch stereo-to-RCA-jack adapter so that it can be fed into the line input of any stereo system.

The cabinet-mounted system shown in Figure 7-7 is great for filling a room with rich, bassy white noise that sounds more natural than what comes from a smaller speaker. Because of the full range the quality speaker allows, all the tones are represented rather than just the higher, "hissy" frequencies. This version of the white-noise generator is great for use as a sleeping aid or when you want to work in a noisy environment without being distracted by background noise.

There are many experiments you can do with white noise, so I will leave you to your own "evil genius" devices to have fun with this project now. Some ideas for design might be the addition of a voltage sweep circuit to simulate ocean waves or perhaps the addition of some type of audio equalizer or filter to shape the white noise into pink or brown noise. Next up, we will dive into the world of lucid dreaming.

Project Eight

Introduction to Lucid Dreaming

Imagine being able to immerse yourself in a world where you can do whatever you like without consequence and also have the ability to alter the very laws of physics and control everything you survey. This may sound like an impossible movie scenario, much like *The Matrix*, or a future prediction as to how virtual reality one day may be able to "jack" directly into our consciousness, but let me assure you that this utopian world can be yours right now without installing any plugs on the back of your neck! Lucid dreaming is an unlikely mix of waking reality and a vivid dream, a dream in which you actually can control your surroundings while aware of the dream yet at the same time experience your dreamscape as if it were the real thing.

The scientific community has been studying and documenting the lucid-dream state since the early 1900s, and the term *lucid dreaming* was first used by the Dutch psychiatrist Frederik Van Eeden. Of course, the mainstream scientific community doubted the very existence of lucid dreaming for many decades because it seemed highly unlikely that a person actually could be conscious and dreaming at the same time. Those who are "natural" lucid dreamers, of course, knew otherwise, but how do you prove that your experiences in the "other world" are true lucid dreams? It's much like trying to prove that there is an afterlife because to be in that state means severing all ties to the real world.

In 1978, internationally renowned psychologist Keith Hearne found a way to create a bridge between the lucid dreamer and the real world in an interesting experiment using a polygraph to measure eye movements. The idea was that on entering a lucid state, the dreamer was able to remember to move his or her eyes back and forth 8 or 10 times in the dream. Knowing that every part of the body with the exception of our eyes becomes paralyzed during sleep, Hearne was able to prove that the lucid dreamer was indeed conscious and able to take control while in a true dream state, as measured by an electroencephalograph (EEG), which measures brain wave activity. While doing this groundbreaking experiment, Hearne also determined that a lucid dream occurs during the rapid-eye-movement (REM) state of sleep. He also discovered that a lucid state occurs most often in the first 30 minutes of the last stage of REM sleep (normally in the early-morning hours). This newly discovered information was very important and set the stage to further understand lucid dreaming and direct many more experiments on the subject.

Knowing when a lucid dream is likely to occur along the typical sleep stages makes it easier to perform experiments in which the subject is told to signal the outside world or in which the experiment involves attempting to induce a lucid-dream state on the subject. The most important and obvious signal that we have is the REM cycle that occurs four or five times during a typical night of sleep. Because a lucid dream is most likely to occur during the last REM cycle, we can exploit this information and create various devices designed to detect this REM period by watching the subject's eyelids and then send some type of signal to the dreamer that it is time to become lucid. Of course, there is more to it than this, so let's have a look at what happens during a typical night of sleep in our gray matter.

There are five stages of sleep, each corresponding to a change in brain wave patterns. Brain wave patterns are measure by an ultrasensitive amplifier known as an *electroencephalograph* (EEG). These brain waves show our current state of awareness and change in five typical steps during a night of sleep. Figure 8-1 shows the stages of sleep as well as the occurrence of rapid-eye movements (REMs) through each stage.

The five stages of sleep are shown in Figure 8-2, with REM sleep being the fifth stage, occurring four to five times during a typical 8-hour sleep cycle. REM sleep in adults typically accounts for about 90 to 120 minutes of a night's sleep during an 8-hour period. REM periods are usually quite short at the beginning of the night, with the final one lasting the longest, where most lucid dreams also occur. Body temperature, heart rate, and breathing are quite irregular during REM sleep, which is another method of detection that can be used to trigger lucid-dream hardware. Of course, the most obvious method is to monitor eyelid movements because this is the most direct indication that a dreamer is experiencing REM sleep. During the various sleep stages, the brain frequencies change quite dramatically (as shown by an EEG), corresponding to different states of awareness. Figure 8-2 shows the various brain waves and their frequencies for each stage of sleep.

Notice in Figure 8-2 how the brain waves and their frequencies change during the five cycles of sleep. Alpha waves are the dominant brain waves while a person is awake and relaxed, which is why they are also present during the initial transition between restfulness and stage 1 sleep. As a person drifts deeper into stage 1 sleep, theta waves become the dominant brain waves. These waves also are associated with a trance-like state or are present when a person is doing some

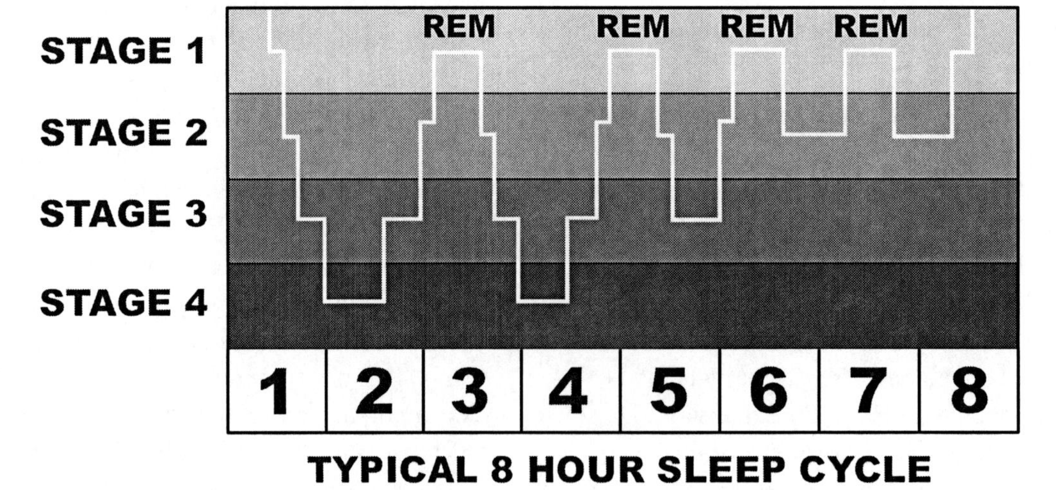

Figure 8-1 *REM frequency during the sleep stages.*

	WAVE TYPE	FREQUENCY	BRAINWAVE SHOWN ON EEG FOR 20 SECONDS
STAGE 1	ALPHA THETA	8-13 Hz	
STAGE 2	THETA SPIKES	12-16 Hz	
STAGE 3	DELTA THETA	0.5-4 Hz	
STAGE 4	DELTA THETA	0.5-2 Hz	
REM	BETA	15-40 Hz	

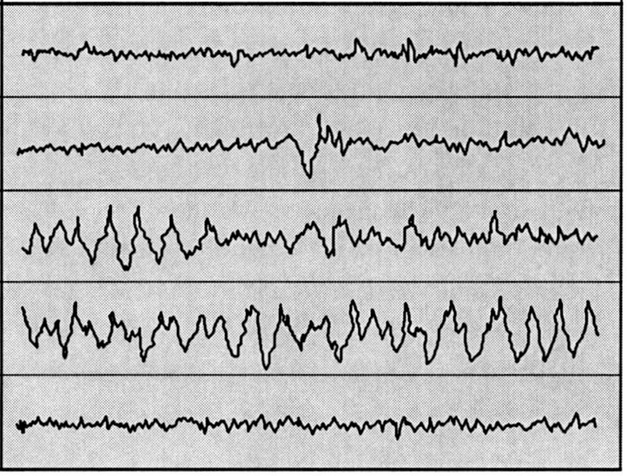

Figure 8-2 *Brain waves during the sleep stages.*

repetitive task where very little thinking is involved. Daydreaming or driving for long distances through a barren landscape most likely would put a person's brain into the theta-wave state. Moving into stage 2, theta waves become dominant and begin to spike and descend at random intervals. The third stage of sleep is considered the onset of deep sleep because theta waves and slower delta waves take over, and wave amplitude is at its highest. Stage 3 of the sleep cycle is where extreme nightmares, bedwetting, or talking out loud occur. Stage 4 sleep is the deepest sleep, with brain waves at their lowest frequency, and this stage happens right before the onset of REM sleep. In stage 4 sleep, more than half the waves are delta waves. In the REM stage of sleep, brain waves become beta waves, which are much the same as when a person is alert or working at some task. Beta waves are produced because the brain is alert and dreaming, yet the entire body is paralyzed with the exception of the eyes. It is thought that without this beta-sleep paralysis, the body might try to act out the dream. It is the REM stage of sleep we are most interested in because this is where you will fall into a lucid dream once you learn to do so.

If you are one of those people who has never experienced a fully lucid dream or don't think you dream very often, don't worry; anyone can learn to lucid dream, and it's likely that you simply forget your dreams as you wake up. Dream recall is the first key to opening the door to the land of lucid dreaming and involves nothing more than keeping a notepad or voice recorder handy so that you can make a short note the second you open your eyes. Have you ever had what you thought was a dreamless night only to have a dream recall triggered by something you saw in the real world that related to the dream? Without that trigger, you would have totally forgotten the dream, but with only something as simple as a visual or auditory cue, you get a total recall. The brain places dream memories in that space that is easily washed away because there really is no point in filling up your memory with random information. I can assure you that if you wake up and write down even one word pertaining to a dream, you will recall many of the details when you look at your note later on. Often, you won't even remember writing in your dream notepad because the brain is in a state that allows memory to slip away in a hurry. A fully lucid dream usually will not disappear from your memory like a typical random or semilucid dream because waking up from your adventure leaves you feeling energized and excited.

Regular sleep patterns also increase the likelihood of a lucid dream because you can use a dual alarm system to wake yourself an hour before your final wake-up time and then drift back to sleep thinking about how much you want to have a lucid dream. This simple technique, along with some of the projects presented later in this chapter, really can increase the number of lucid dreams you have once you learn to deal with them.

Much like riding a bicycle for the first time, your first lucid-dream encounter probably will be short and send you crashing into the land of wakefulness. The instant you realize that you can have total control, you may wake up because of the excitement, or you may not "trust" your dream world enough and scare yourself awake. I remember this happening when I first started studying lucid dreaming and was constantly trying to fly. The first few times I became aware, I woke right up feeling a huge sense of accomplishment and excitement, but that, of course, killed the dream right away. Once I learned to calm down and "get with the program," I often would try things like jumping into the air or changing my surroundings, which also caused a fast awakening. Because your subconscious knows that flying through the air without wings or breathing underwater is something that obviously would be dangerous, your untrained lucid-dreaming mind often makes you feel unsure of the experience and causes an abrupt end to your adventure. In time, these things also can be overcome.

One of the great lucid-dream researchers of our time, Stephen LaBerge, a Ph.D. in psychophysiology, has written many good books on the subject and has come up with many useful techniques and even devices that can be used to enhance your lucid-dreaming experience. If you would like to expand your knowledge on the subject of lucid dreaming, just Google his name, and you will find more than enough information on his research, books, and inventions, some of which gave me ideas for a few of the projects in this chapter. Along with much groundbreaking research on lucid dreaming, LaBerge has come up with some very simple techniques for enhancing the chance of a lucid dream, as well as ways to prolong the experience.

One of LaBerge's best-known methods for enhancing the chance of a lucid dream is call the *MILD technique*, which stands for *m*nemonic *i*nduction of *l*ucid *d*reaming and is basically a way of reminding yourself to have a lucid dream. This is done by waking yourself before your normal wake-up time and becoming fully alert by walking around for a few minutes, reminding yourself that you are now going back to a lucid dream. You then crawl back into bed, thinking about the dream you plan to have or recalling the one you were having just as you woke. You must *will* yourself to go back into a dream and become lucid. Surprisingly, this system may work better than most of the electronic devices used to attempt a lucid-dream induction, and it requires only a clock with two alarms or a pair of alarm clocks.

Another simple approach that will help to "train" your brain to recognize a dreaming state is by *testing reality*. Now, this may seem crazy, but often a dreamer simply will forget that it is not normal to be traveling through space in a ship made of cardboard boxes, or the dreamer simply will accept the fact that the dog now talks and drives a motorcycle. By testing reality while you are awake, you get into a pattern that will follow you into the dream world, and when it does, you will begin to realize that reality is bent. Reality testing may be as simple as holding up your hand every few hours to count your fingers because often you cannot do this in a dream or there are an odd number of fingers. You also may ask yourself routinely, "Is this a dream?" during the day because if you do this in an actual dream, your mind will notice things that don't belong. The key to reality testing is repetition, so I have included in this chapter a simple reality tester that will send you a reminder once every hour or so to perform a simple reality test. When you get used to the device, it will begin to appear in your dreams and could be the key to realizing you are in the "other world."

Some of the other techniques that will be explored as projects in this chapter involve presenting some external stimulus to the dreamer either at opportune times based on sleep patterns or via feedback directly from the body. Using audio to direct a dream is interesting because your mind is often aware of what you hear while dreaming, and if you have ever fallen asleep while watching a movie, you might have had the experience of remembering a dream that had to do with whatever you were listening to at the time. Mixed with lucid dreaming, this approach can become very interesting, especially if you create some type of audio dreamscape and use your own voice to "coach" your dreaming self into a new lucid-dream world. One of the projects in this chapter also will respond to movements of your eyelids, so you can use this feedback in conjunction with a clock to determine when you are in the last REM stage of sleep before waking because this is the opportune time to introduce some type of audio or visual cue to help trigger a lucid dream.

Well, that's enough hype about lucid dreaming for now! Honestly, who wouldn't want to have a virtual reality without limits to control every night (or at least once in a while)? No computer game or thrill-seeking adventure could ever compare to a lucid dream because there are dangers, moral issues, and of course, the very fabric of reality to contend with. When you wake up from a full-blown lucid dream, it won't feel the same as when you wake up from a typical strange dream and just think, "Huh?" Your lucid-dream experience will seem so real at the time that when you finally open your eyes, it will seem like it really happened, and you will remember the experience much like any incredible event in your life. Chances are that you won't be able do lucid dreaming often, so don't worry; it's not like you are going to get addicted to the sleeping world and leave your waking life behind!

Now let's dig into the electronics junk box and put our "evil genius" minds to work so that we can master the other half of our lives—the dream world.

Project Nine

Waking-Reality Tester

Now here is a device you probably thought you would never need—a little box that helps you to determine if you are awake or dreaming. This may sound like a completely useless gadget, but when it comes to your lucid-dream arsenal, this is actually a very useful tool that can greatly increase your chances and the frequency at which you might have a lucid dream. By itself, the device serves no real function besides being a simple timer that reminds you to practice one of the best-known techniques used to aid in lucid dreaming. Reality testing on a regular basis becomes a habit that is often transferred into your dreaming world, where, of course, reality is much different from that of the waking world. When you ask yourself, "Is this real?" in a dream, often you realize that you are in fact dreaming and then can will yourself into a fully lucid state. This system only works if practiced regularly, so this little timer helps you to get into the habit of testing your current state of reality in order to bring this habit into your dream world.

The reality tester is a covert device that works much like a pager, so you can carry it at all times without having to explain your "weirdness" to those around you. Since the reality tester simply vibrates every hour or so, you can continuously test your state of consciousness without interrupting your day. Eventually, you will begin to do this in your dreams, and when you look around and ask yourself, "Is this the real world?" you will often realize that things are just too strange to be real and then can become lucid. A tried-and-true method of reality testing is to simply count your fingers because often in a dream there will be too many, too few, or you simply will not be able to do it at all. Reading some text is also a good method of reality testing because the dream world often will return scrambled text or something that looks more like ancient hieroglyphics. In a dream, some things will just be plain ridiculous, such as a purple sky or the fact that you are eating your breakfast on a pirate ship made of Lego blocks. In this case, there is no doubt you are dreaming, but you may not even question these oddities unless you are reminded to do so by the reality tester. Once you get in the habit of realizing your shift in reality, you can begin to take control over your dream world and rearrange it as you see fit.

The reality test can be made to vibrate covertly, or it can be adapted to make an audio tone (or both if you like). The main schematic for the reality tester is shown in Figure 9-1 with a small dc motor used as a vibration device.

The 555 timer (IC1) is wired as an adjustable oscillator that sends its pulses to a 74HC4040 14-bit binary counter (IC2) so that the pulse time

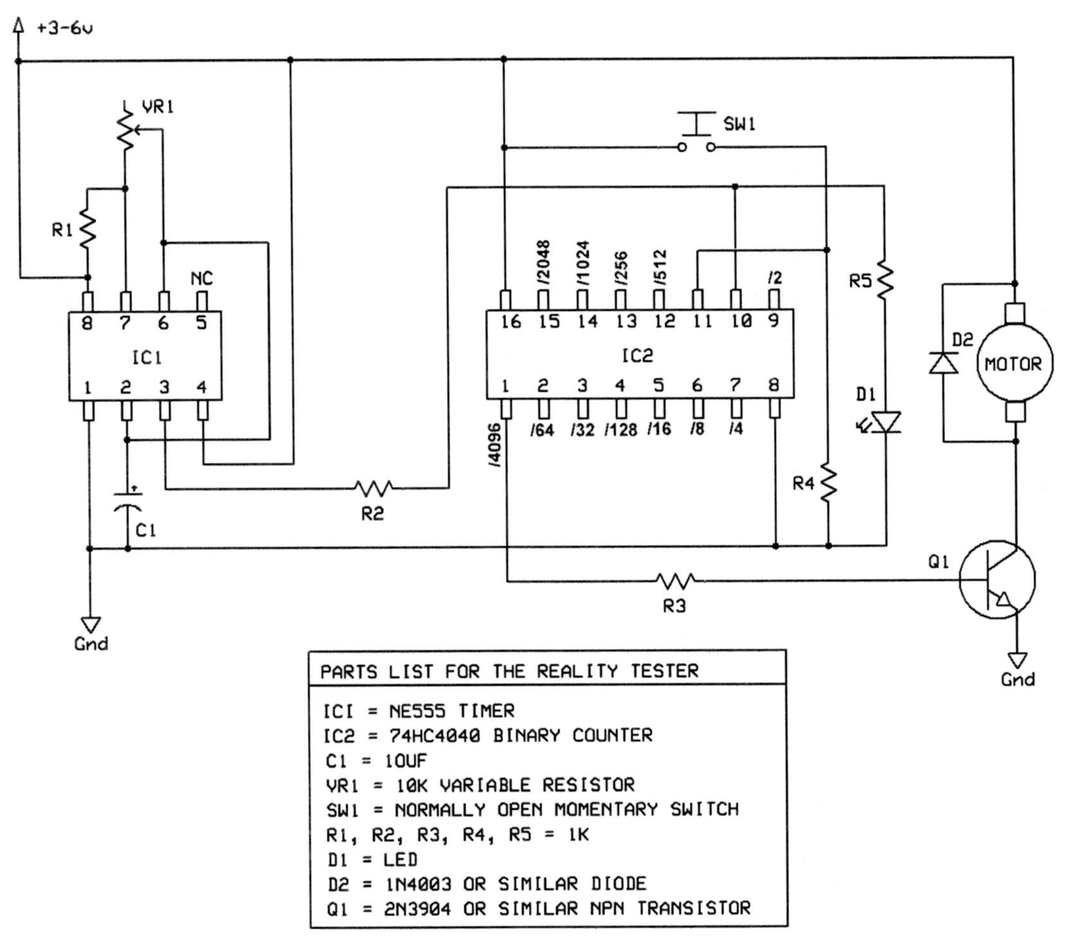

Figure 9-1 *Reality-tester schematic with vibration motor.*

can be divided by one of the following values: 2, 4, 8, 16, 32, 64, 128, 256, 512, 1024, 2048, or 4096. Using the 4040 to divide the pulse time makes it much easier to set up a timer based on minutes or hours because the 555 timer is not very good at outputting long delays. In the schematic, R5 and D1 are used only during setup so that you can use a clock or stopwatch to count the timer pulses and then adjust VR1 to get the light-emitting diode (LED) to flash at about once per second. Once you have a 1-second flash rate, just choose one of the outputs from the 4040 counter to set the interval at which the reality tester will buzz you. In the schematic shown in Figure 9-1, I have chosen pin 1 of the 4040 counter, which is marked 4096, so my reality tester will alert me about once every hour if I set the 555 timer to a pulse rate of about 1 second. To calculate your final time, use the formula

4096/60 (seconds)/60 (minutes) = 1.13 hours

Transistor Q1 is being used as a switch to turn on a motor that operates as a vibration device because it has an unbalanced counterweight on its shaft (more on this later). Switch SW1 is a pushbutton switch that forces you to reset the unit to acknowledge the fact that it is time to test your reality. The reality tester also can be made to output an audio tone by removing the motor, Q1, and R3 and replacing them with the circuit shown in Figure 9-2. As you can see, this circuit is much like the 555 timer clock oscillator in the main schematic except that it oscillates at a higher frequency and then sends its output into a piezo buzzer.

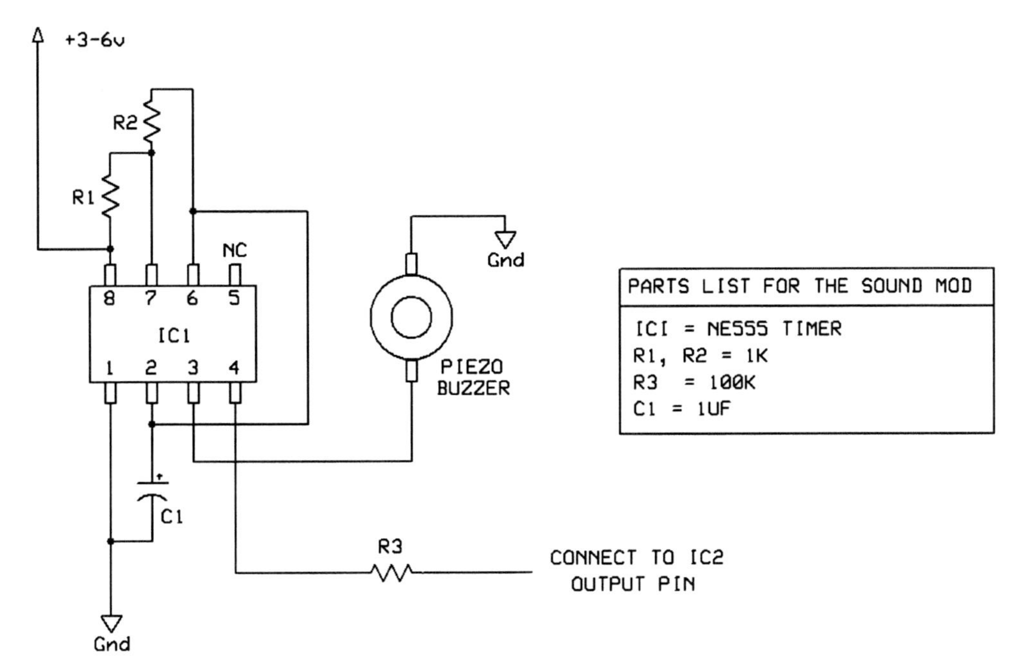

Figure 9-2 *Reality tester audio output modification.*

You could have both the vibrating motor and audio at the same time or add a switch to change between audio and vibrate modes. Having both modes of operation might be good because it is more convenient to use the audio alert mode when working alone at a desk and less embarrassing to use the covert buzz mode when out and about. When you pull a beeping home-built gadget out of your pocket at dinner and try to explain to people that it helps you keep track of your current reality, that might get you some very strange looks!

Now let's work on the mechanical part of the project that will create a vibration very similar to a cell phone or pager. Figure 9-3 shows a number of small dc motors that can be used to create a vibration unit, as well as a few that are actually taken from pagers and cell phone. The two small motors shown at the bottom left of the figure are actual vibration motors, as you can see by the unbalanced counterweights already attached to their output shafts.

Don't worry about tracking down a pager motor because it is extremely easy to make the same kind of thing with any small dc motor and a soldering iron. All you need is a small dc motor (the smaller, the better) and a blob of solder stuck to the end that will cause the motor to vibrate as it spins. These motors can be found in all kinds of kids' toys, CD-ROM or DVD players, video cameras, electric shavers, and many other small appliances. Don't worry about the voltage of the motor, just that it spins up when you place it on your 3- or 6-V battery. Even a motor rated for 24 V probably will spin fast enough on 3 V to create a vibration, but you really don't need a huge motor because the device should be small enough to fit in your pocket when completed. To add an unbalanced counterweight to your motor's shaft, just heat up a blob of solder, as shown in Figure 9-4, and then dip the motor shaft into it and hold it there until the solder solidifies.

As the solder blob cools, try to keep the motor shaft perfectly still so that the blob stays tight on the shaft. When the solder has cooled, add power to your motor and see how it vibrates. Chances are good that the solder blob will not be balanced perfectly on the shaft and that there will be a lot of vibration as the motor spins. Another method of making a counterweight is to find a gear or

Figure 9-3 *Small dc motors that can make vibrations.*

Figure 9-4 *Adding an unbalanced weight to the motor.*

Figure 9-5 *Motor with counterweight attached.*

wheel that fits onto the shaft and file one side away. You will need to solder wires to the power pins on your motor, as shown in Figure 9-5, so that you can connect it to your breadboard and then to your completed circuit.

The breadboarded circuit is shown in Figure 9-6, and you can get a perspective on how small my vibration motor really is compared with the other parts. The 3-V lithium battery will power the 5-V logic chip and timer without any problem and give plenty of juice to the motor to make it vibrate.

The circuit should be built first on a solderless breadboard so that you can adjust the timer properly using the temporary flashing LED (shown by the arrows in Figure 9-6). By adjusting the variable resistor (VR1), you can set the flash rate to about once per second and then "tap" into the 4040 counter to set the overall delay between reminders. Just take the divider number (shown on the pins of the 4040 counter in Figure 9-1) and divide that number by 3600 to get the approximate value in hours between reminders. Once you are satisfied with the time between reminders, you can remove R5 and the LED so that the LED doesn't waste battery power. You also can measure the resistance across the variable resistor once it is removed from your circuit and then replace it with a fixed-value resistor of the same value. Once the circuit is working properly, the components can be transferred to a small perf board for installation into a portable cabinet.

As shown in Figure 9-7, I also added an on/off switch to the system so that battery power could

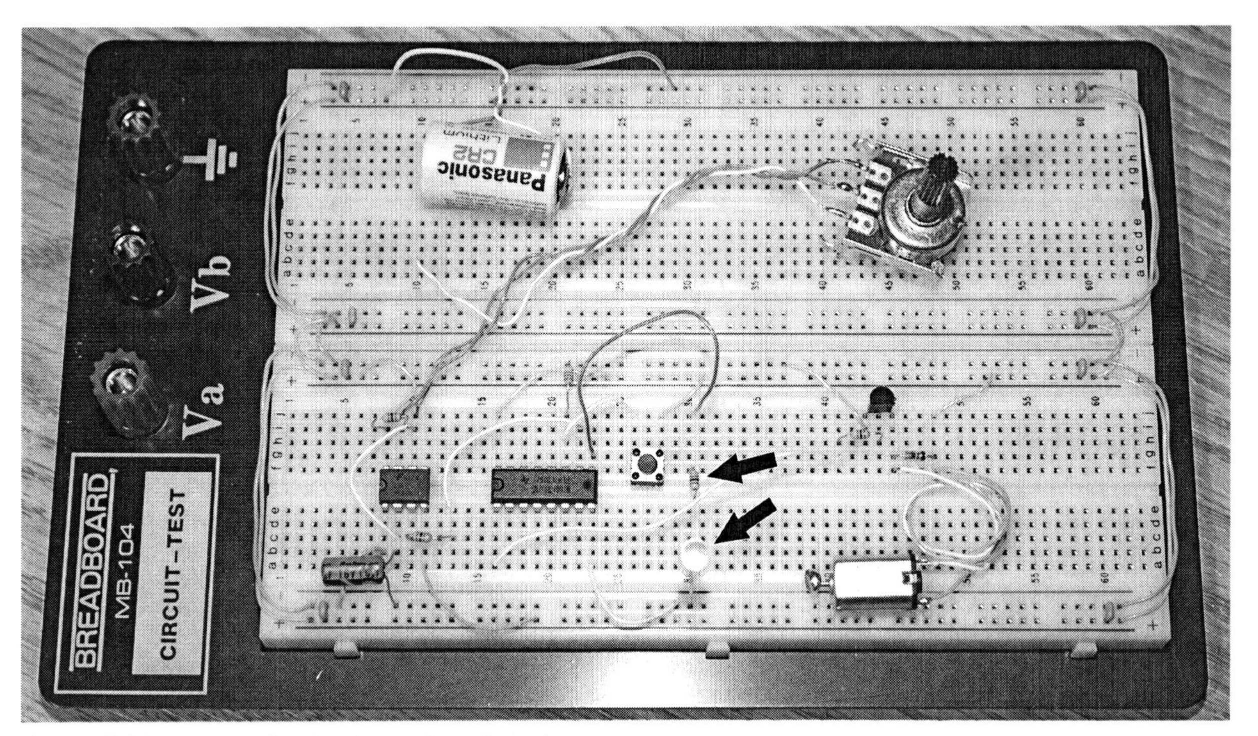

Figure 9-6 *Testing the circuit on a breadboard.*

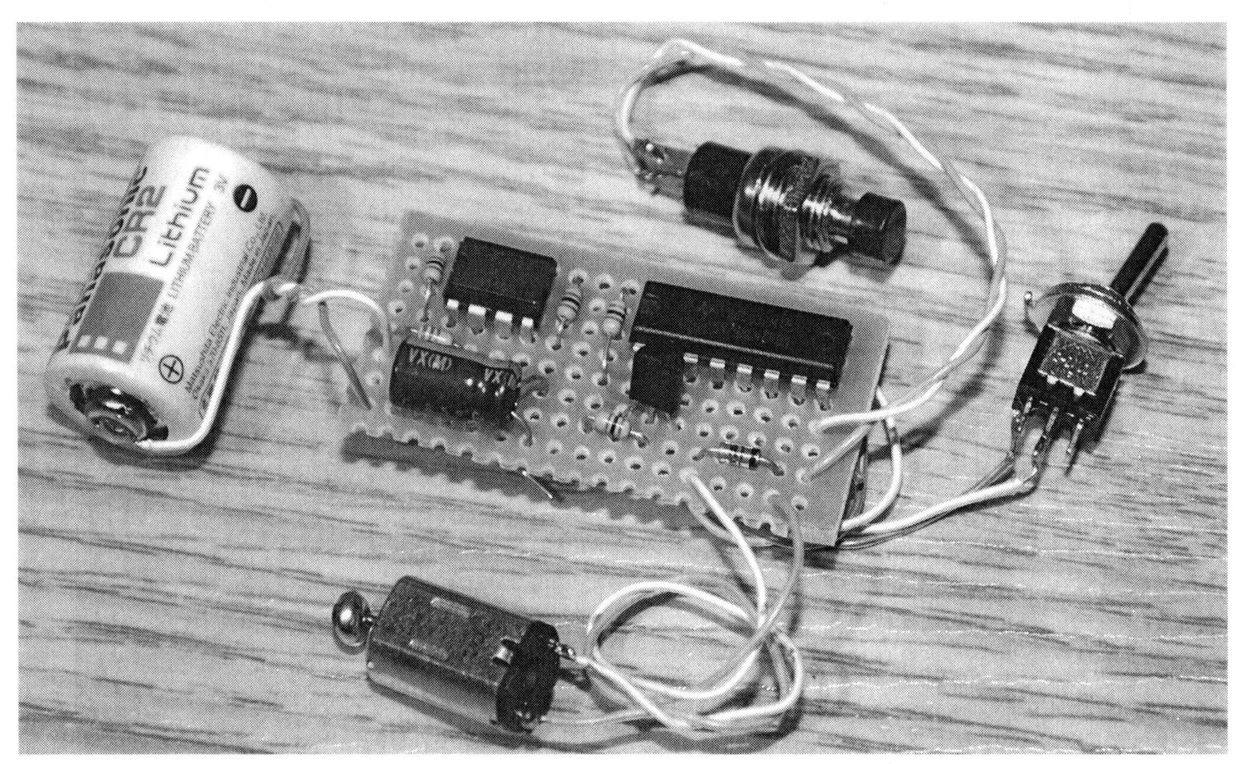

Figure 9-7 *A completed circuit on a perf board.*

be conserved when not busy testing reality. Now all you have to do is find the smallest possible cabinet for your reality tester, and you can begin training your mind to regularly question your current reality. I found a plastic film container that was just perfect to fit the small battery, motor, and circuit board inside, and it fits nicely in a pocket without becoming too annoying. Figure 9-8 shows the tight fit of all the components into the film container, with just enough room for the reset button out the top.

Before I jammed the works into the film container, I wrapped the circuit board with tape so that the battery case and on/off switch would not short out against the component pins. This is not shown in Figure 9-8 because it looked like a mess, but you should consider wrapping your perf board in tape if it is to be jammed into a small container such as this. I also decided to place the on/off switch in the container because I did not want it turned off accidentally when I was carrying the unit around. The reset switch is not a problem because it only works once the buzzer or vibration has already begun. Now you have the lengthy process of testing the unit with it fully assembled. It is painful to wait for the entire cycle, so find something to do and then flip the switch, keeping track of the time you turned on the unit. If you were shooting for an hour delay, then your buzzer or vibration should begin somewhere close to that time, give or take 15 minutes for timer inaccuracies.

The completed reality tester is shown in Figure 9-9, ready to carry around. Of course, I do not recommend that you attempt to bring this device on a plane with you because it surely looks like something dangerous, and trying to explain its purpose to security folks might make you look like a crazy person! Every time you are alerted by the reality tester, press the reset button and then ask yourself, "Is this a dream?" Look around and try to find some text to read, or hold out your hand and count your fingers. In the dreaming world, text almost always will be unreadable, and your fingers will not add up correctly or just look strange somehow. It may take a week or more of

Figure 9-8 *Everything fits into a small film container.*

Figure 9-9 *The reality tester ready for action.*

constant reality checking before this habit spills over into your dreams, but once it does, it will be easy to trigger a lucid-dream state by realizing that you are actually dreaming. You also can expect a few misfires as well as you get excited and wake up too quickly at the prospect of actually controlling the dream. Like all things worthwhile, lucid dreaming is an art that must be practiced to be fully appreciated.

Once you've managed to learn the art of realizing that you are dreaming, you might want to try some other experiments on your sleeping brain. The next project will help you to "direct" your dreams by audio suggestions presented at opportune times during your sleep cycle. Some things that you might want to try are sleep learning, subliminal suggestions, movie recordings, random sounds, and motivational material.

Project Ten

Audio Dream Director

Have you ever fallen asleep while watching a movie, only to awaken a short time later and recall a dream that had something to do with what you were watching? How about waking up early because of some neighborhood distraction, only to realize that you were dreaming about something that was making a very similar sound? Much like a waking mind can react to subliminal messages, the dreaming mind also can be influenced by the senses, sometimes to a greater degree. To exploit this fact, we will build a device that will send you an audio message at the opportune time during your REM sleep phase (about an hour before you wake). This is the time when you are most likely to enter a lucid dream and the best time to remember a dream or make notes for later recall.

The idea behind this device is quite simple. A hacked alarm clock is modified to trigger some type of audio player that will send you a prerecorded audio file to help you enter a lucid dream or simply "direct" your dreamscape in some direction. There is no direct feedback from the body in this version of the device, but since we know that the last hour of sleep is usually the most important for dream recall and lucid dreaming, the trigger is simply set to an hour before your real wake-up time. You can then prerecord 20 to 40 minutes of audio for playback into a pillow speaker and let your mind wander in the direction of the audio clip. Maybe you want to learn Spanish while you sleep? How about recording your favorite movie and trying to live out the action? Maybe you just want to record yourself speaking about how tonight you will enter a lucid dream? Chances are the audio will affect your dream, although it may do so in ways you did not expect.

When I first experimented with this device, I found that the audio certainly did affect my dreams, but sometimes in ways I was not expecting. I once recorded the unmistakable sounds of a car race, thinking that I would wake from a dream where I was sitting behind the wheel of an Indy car. When I did wake and had dream recall, I remember that the dream was about some weird tiny flying robot, and the car engine sounds actually were the sounds of the little wings flapping as the robots flew past my ears. Although this was not what I expected, it certainly had an effect on my dreams. I also tried my own voice talking to me like some guru about how I was entering a lucid dream, and this worked a few times. On other occasions, I had dreams about making the actual audio file!

The dream director is just an amplified switch that locks down a relay when there is a small change in voltage at the input. The input typically is the output buzzer from an alarm clock. The relay switch then is connected to the play button on some type of audio playback device such as an MP3 player, handheld audio recorder, or even a CD player. Since the actual alarm speaker or buzzer is removed from the hacked clock, you do not wake, and the audio file begins to play through a pillow speaker under your head. Using any computer sound program, you can compile your audio file to slowly ramp up the volume so that you are not awakened suddenly by the sound. If all goes well, your dreaming mind will drift into a dreamscape influenced by the audio file. Because you may remember making the file or have your voice dubbed in to remind you that this is a lucid dream, chances are good that something interesting will happen.

Figure 10-1 shows how simple the relay-trigger part of the dream director really is. IC1 is an OR gate connected in such a way that its own output feeds back into its input and latches it in the ON position once triggered by an external voltage change from your alarm system. The output then is fed into transistor Q1 to drive the relay coil and close the contacts. Using a relay isolates your device from the rest of the circuit and makes it easy to hack just about any audio player with a play button that will start the audio playing. The relay also adds complete isolation from the alarm clock, although I still would recommend that you only use battery-powered devices when "jacking" yourself up at night to your hacked gadgets. Just think about lightning storms, and you will realize why not being plugged into the wall socket at night is a good idea.

You can run the circuit on from 3 to 6 V, and just about any relay with a voltage rating from 5 to 12 V will work. Even some relays rated at 24 V seemed to latch just fine on my system when I was running it on only 3 V. The best way to find a suitable relay is to try activating it with whatever power source you intend to use here. You do not need a large relay because current is very minimal, and your relay needs only a single

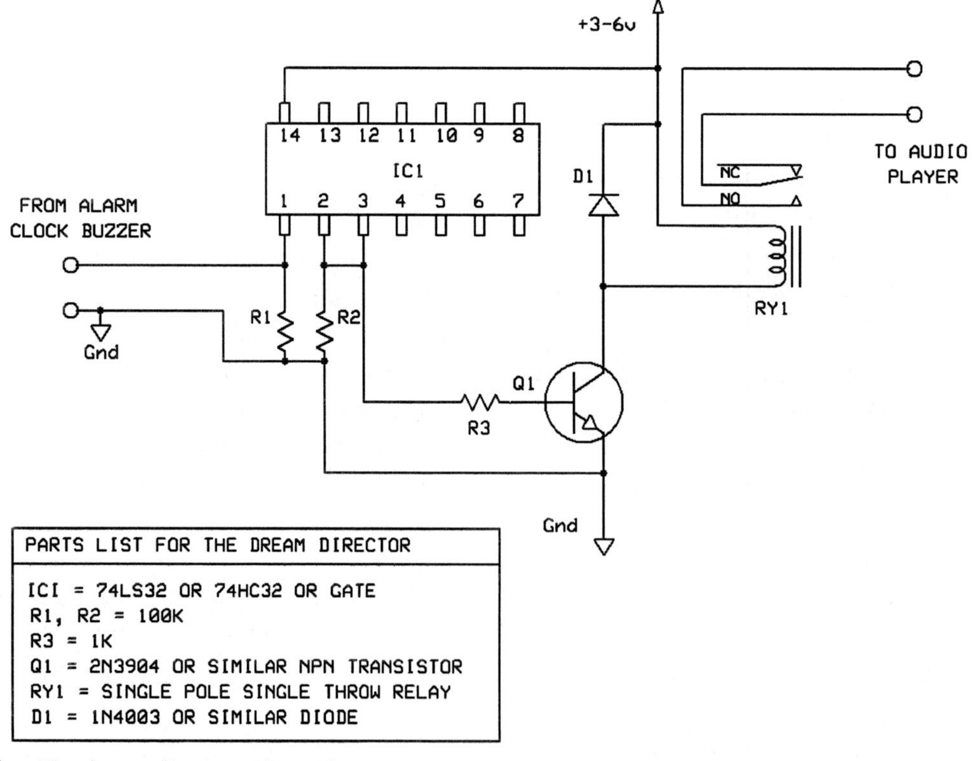

Figure 10-1 *The dream director schematic.*

pole with a pair of normally open contacts. Some of the relays I tested on my 5-V battery are shown in Figure 10-2, so I chose the smallest one, which was taken from an old modem card and rated for 5 V.

This circuit is so simple that you may not even need to test it on a breadboard, but it is certainly easier to make changes this way in case something does not work the way you want it to. Figure 10-3 shows the relay driver ready for testing once I rip the buzzer from my alarm clock and feed it into the logic gate. The reason the circuit latched the relay closed is because once your alarm clock starts feeding the input, it will do so intermittently, which may cause some audio players simply to start playing the same audio file over and over again. By latching the relay, the play signal is sent to your audio device only once.

The completed relay driver is shown built on a small perf board in Figure 10-4. Notice the 5-V regulator installed in the center of the board so that I can run the driver from a 9-V battery. This 7805 regulator is shown in Figure 2-8 and will output a perfect 5 V from a supply of 9 to 12 V. Of course, 3 to 6 V would have been just fine for the logic chip, but 9 V most likely would cook it in a hurry, so the regulator is necessary. When the relay driver is working, any small change in the input voltage will cause the relay to close and not open until the power is removed. To test the board, just connect a 1.5- or 3-V battery to the input wires and listen for the relay to close.

The alarm clock that will trigger the relay board can be just about any type that has a digital alarm connected to a speaker or buzzer with two wires. Figure 10-5 shows two of the clocks I tested, and both worked perfectly with the relay board after removing their buzzer output wires and feeding them to the relay-board input. A battery-operated clock is certainly more preferable than one that will connect your sleep laboratory to the power lines, so take this into consideration when looking for a clock to hack.

Figure 10-2 *Some relays that will work with this circuit.*

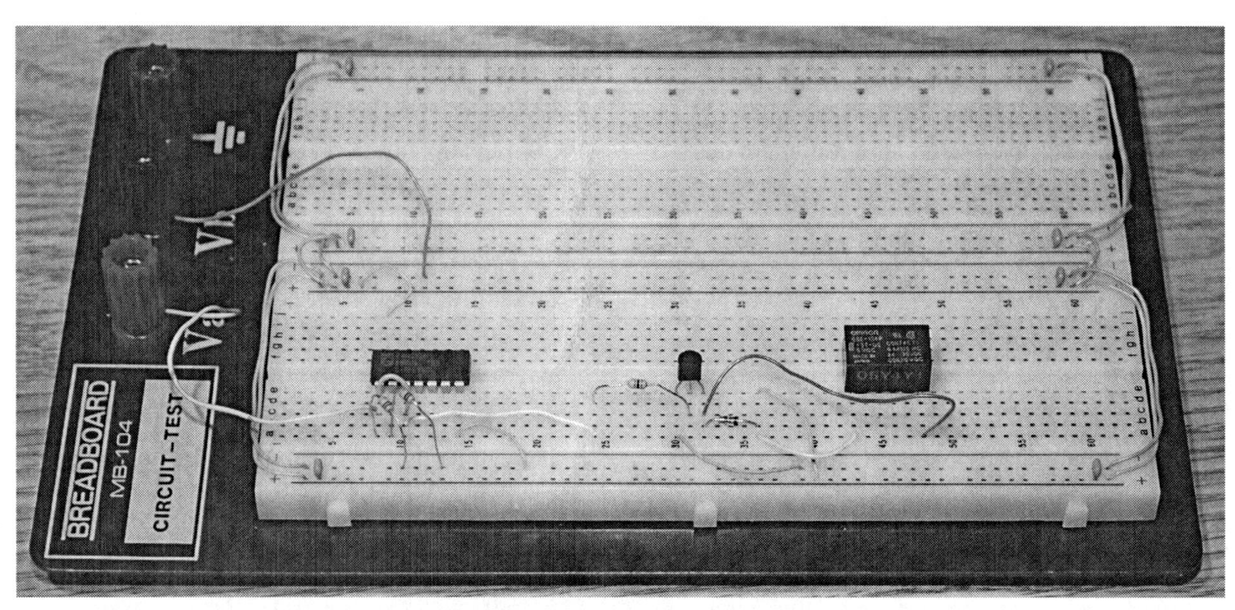

Figure 10-3 *Testing the relay driver on a breadboard.*

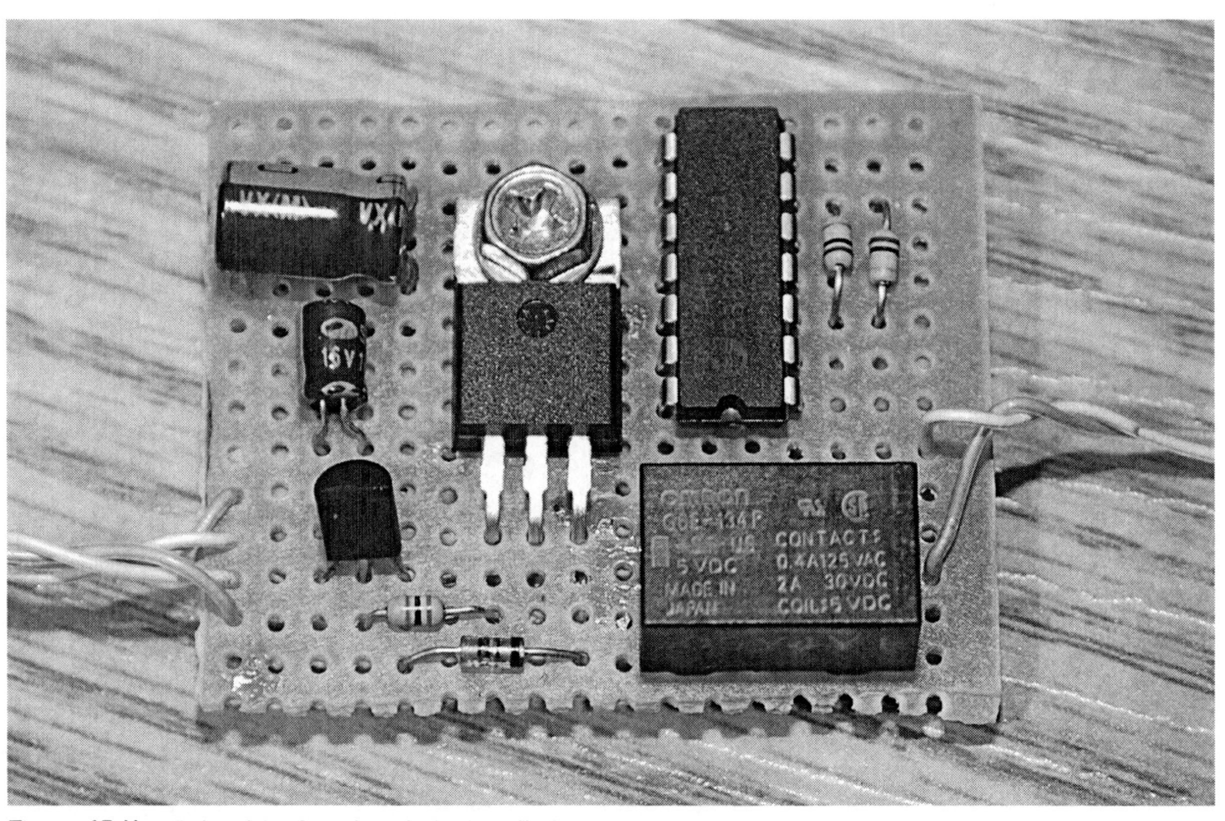

Figure 10-4 *Relay drive board ready for installation.*

Figure 10-5 *Choosing an alarm clock to drive the relay board.*

To trigger the relay board, there needs to be a small voltage change at the input of the logic gate, and this is taken from the output of the alarm clock. You need to identify the speaker, buzzer, or piezo element and remove the two wires from it so that you can install some type of output jack onto the alarm clock's cabinet. Since you don't want the alarm to wake you up, the buzzer must be completely removed from the circuit so that any alarm sounds will be fed silently directly into your relay-board trigger input. Figure 10-6 shows the piezo element found inside the small alarm clock that I pulled apart for this experiment. The two arrows point to the alarm output wires, and

Figure 10-6 *Identifying the alarm output wires.*

there is a positive (red) and negative (black) wire in this unit, which is important because the input to the gate should be a positive-going pulse. Don't worry if your wires are the same color and there is no clear indication as to which one is positive because you can just try them both ways until you hear the relay close when the alarm is going off.

Once you have identified the alarm output wires, connect them to some kind of jack that can be installed in the alarm clock cabinet so that you can plug your relay board or some other experiment into your modified alarm system. Figure 10-7 shows the $\frac{1}{8}$-inch mono jack I decided to use for this device, keeping in mind the positive wire going to the center pin on the jack. Again, if you are unsure about the polarity, just guess, and then reverse the wires if the relay fails to close when the alarm is going off. Another idea for testing is to remove the original alarm buzzer or speaker so that you can plug it back into the newly installed output jack. This gives you the ability to hear the alarm for testing purposes or use the clock normally when not experimenting with it.

The newly modified "silent alarm" system is shown in Figure 10-8 with the alarm output jack installed in the side of the cabinet. This little hack is great for a variety of experiments that might require an accurate timer with an output capable of driving a digital circuit or relay into action. By using this alarm set to trigger your dream experiments an hour before your real alarm sounds, you are almost guaranteed to be in the last stage of REM sleep, where lucid dreaming is most likely to occur. Also shown in the figure is the matching male $\frac{1}{8}$-inch mono plug that will transfer the alarm voltage into the relay-board trigger input. Now you need to install your relay-board trigger in a cabinet or attempt to jam it into the alarm clock cabinet and leach power from the clock.

Figure 10-7 *Adding an alarm output jack.*

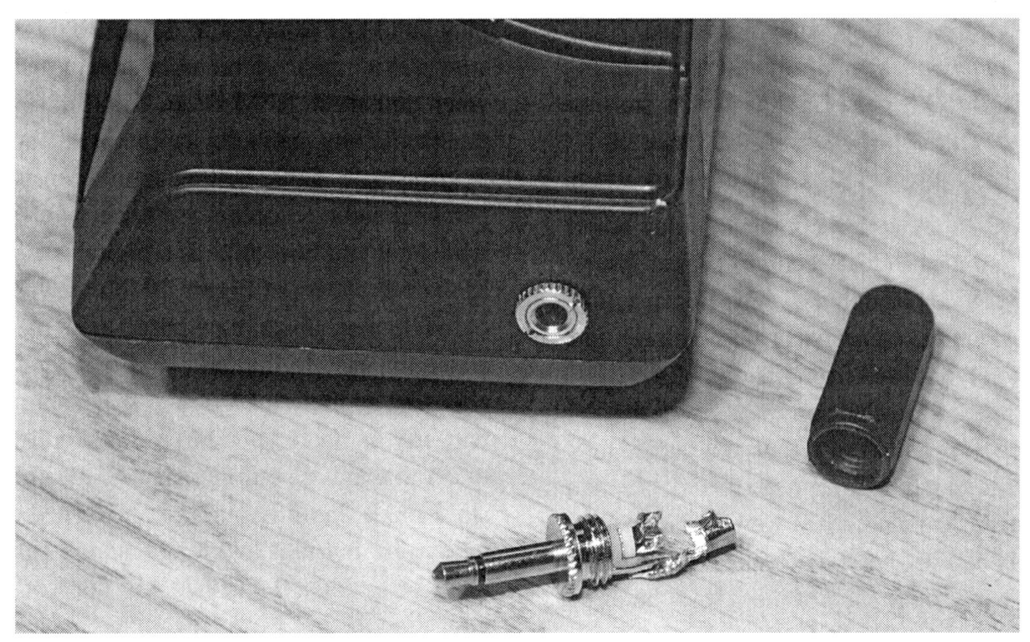

Figure 10-8 *A universal "silent alarm" for your projects.*

I actually had room in my alarm clock for the relay board and found a 12-V dc output from the power supply, so I could have installed everything right in the clock and fed the 12-V supply into the 5-V regulator in the relay board. Of course, this makes the unit unavailable for other experimentation, so I opted for the typical "black box" installation shown in Figure 10-9, making sure that there was also room for a 9-V battery. I also added an LED to show that the unit was on

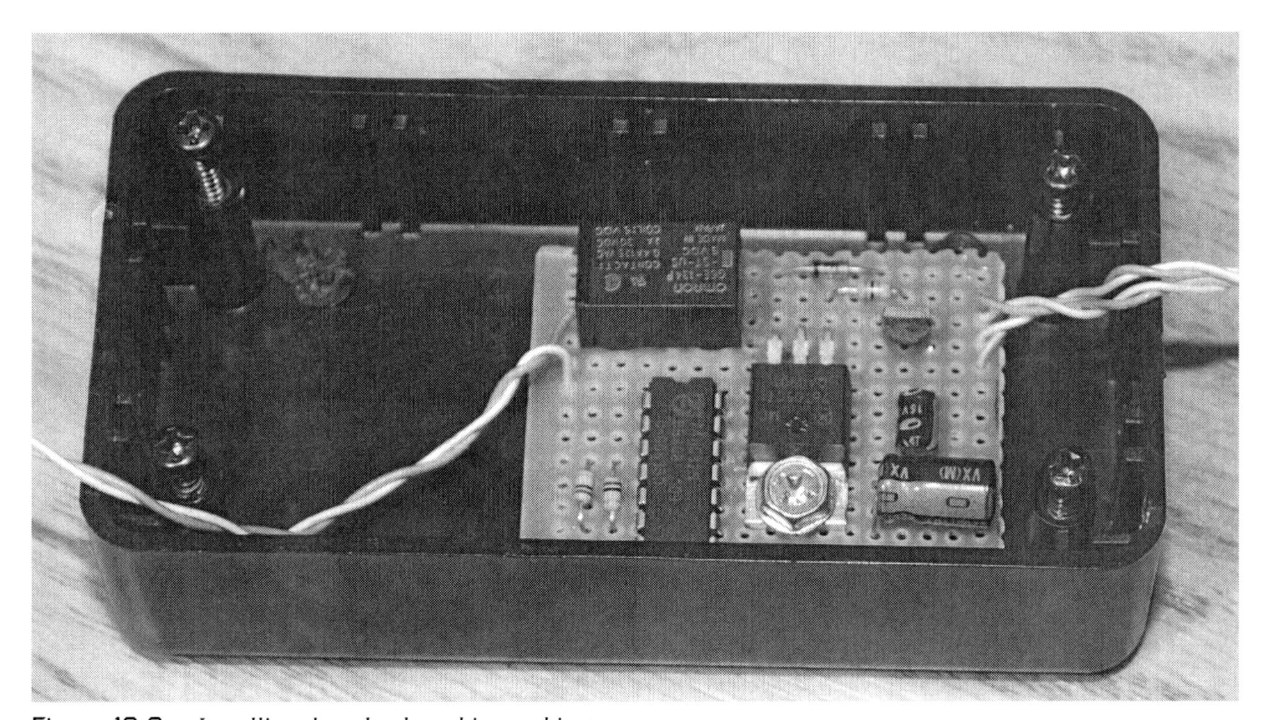

Figure 10-9 *Installing the relay board in a cabinet.*

and a switch to turn it off when not in use. Because the device is on all night, I used a high-value resistor to keep the LED as dim as possible to reduce drain on the battery. A 47K resistor made the LED just barely visible in a dark room.

Figure 10-10 shows the completed relay board and modified alarm clock ready to trigger just about any device through the relay. Because the relay's mechanical contacts carry no voltage and are completely isolated from the rest of the circuit, you are protected somewhat from the alternating-current (ac) supply if you decide to use a plug-in clock, and your external devices are also safe from the circuit if it were to fail because the relay is nothing but a mechanical switch. The RCA-style jack on the top of the box is the direct connection to the relay contact switch so that I can plug devices into the box. I wanted a different jack than the one used to input the alarm so that I wouldn't accidentally mix them up and possibly harm the external devices I plan to plug into the unit.

With your alarm feeding the relay board, all you need to do is connect the relay switch to the play button of some audio device so that it will start playing your prerecorded dream messages when you are in REM sleep. You can hack practically any audio device that will start playing by pressing a single button, so inexpensive personal audio recorders are perfect for the job. I will show you how to hack a digital recorder such as the one shown in Figure 10-11, as well as the old-style mechanical tape-based units that are also quite easy to find and modify. When choosing an audio player, just make sure that it will stay on and respond to a single button press to begin playing your message. Some units have a sleep mode, but often this can be turned off as an option. If your unit will not respond to a single play button, then the device will not work with your relay board. Mechanical tape players always will work because the relay board will control the voltage to the motor while the play button is stuck in the play position. MP3 players usually work as long as they start with a single button press. Usually, the cheaper and larger the unit, the better it will be for hacking.

Figure 10-10 *Alarm clock and relay board ready to use.*

Figure 10-11 *A digital voice recorder is perfect for this project.*

To connect the audio device to your relay board, all you need to do is open the case, find the play-button points, and add two wires back to your relay contacts. Polarity is not important because the relay acts exactly like the original switch on the audio device, and if you have room, you can feed the wires out of the case and not even interfere with the operation of the play button. Figure 10-12 shows how I was able to tap into the play button and sneak the thin wire outside the unit without messing up the original play-button functionality. If you are hacking a very small MP3 player, then you might want to have a magnifying glass handy and use the smallest wire you can find because the contact points will be very small. If you don't care about using the audio device normally, then you can make an ugly hack if necessary, but remember

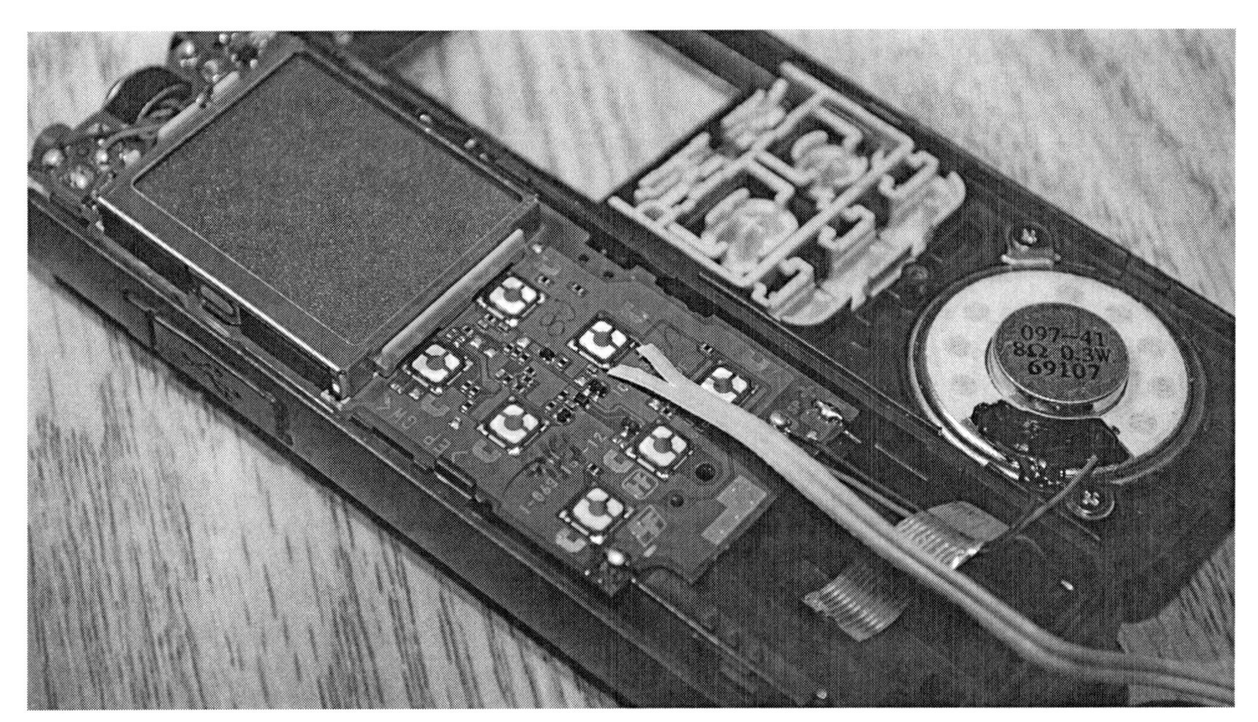

Figure 10-12 *Hacking into the play button.*

that the device will have to be somewhat operational to get your messages into the memory.

Mechanical tape players are very easy to hack and are always guaranteed to work with the relay board because the drive-motor connection will be severed and routed through the relay switch. Open the unit and identify the motor, as shown by the arrow in Figure 10-13. The motor will have only two wires coming from it, and you only need to cut one of them and then feed a pair of wires from each end back to the relay switch. Now you can press play on the recorder and nothing will happen until your relay closes and starts the motor spinning—it's an easy hack! If there is a jack on your recorder with the label "REM" on it, then you are in luck because that's exactly what this jack does: It breaks the connection to the motor so that you can just find a compatible plug and send that directly to your relay switch. Mechanical recorders are nice because they can be found at many stores relatively cheaply and are always easy to hack.

Sound quality is not important here, so you probably can use the same tape over and over again without any problems.

Figure 10-14 shows the completed hack done to the mechanical tape player in order to feed the motor output back into the relay board. I also took this opportunity to add a connection from the original speaker out to another jack so that I could use a pillow speaker rather than placing the device under my pillow directly. This modification reduces the risk of pressing stop or some other button during sleep. Most players already have a headphone jack, so you only have to make the audio-output modification if there is no headphone or external speaker plug on your unit.

Once your relay board is triggering your audio device, you are almost ready to experiment with audio dream induction. The completed unit is shown in Figure 10-15, ready with my favorite movie scenes to influence my dreams. The only question left is, How do you get the audio to your ears in a comfortable way without waking

Figure 10-13 *Hacking a mechanical tape player.*

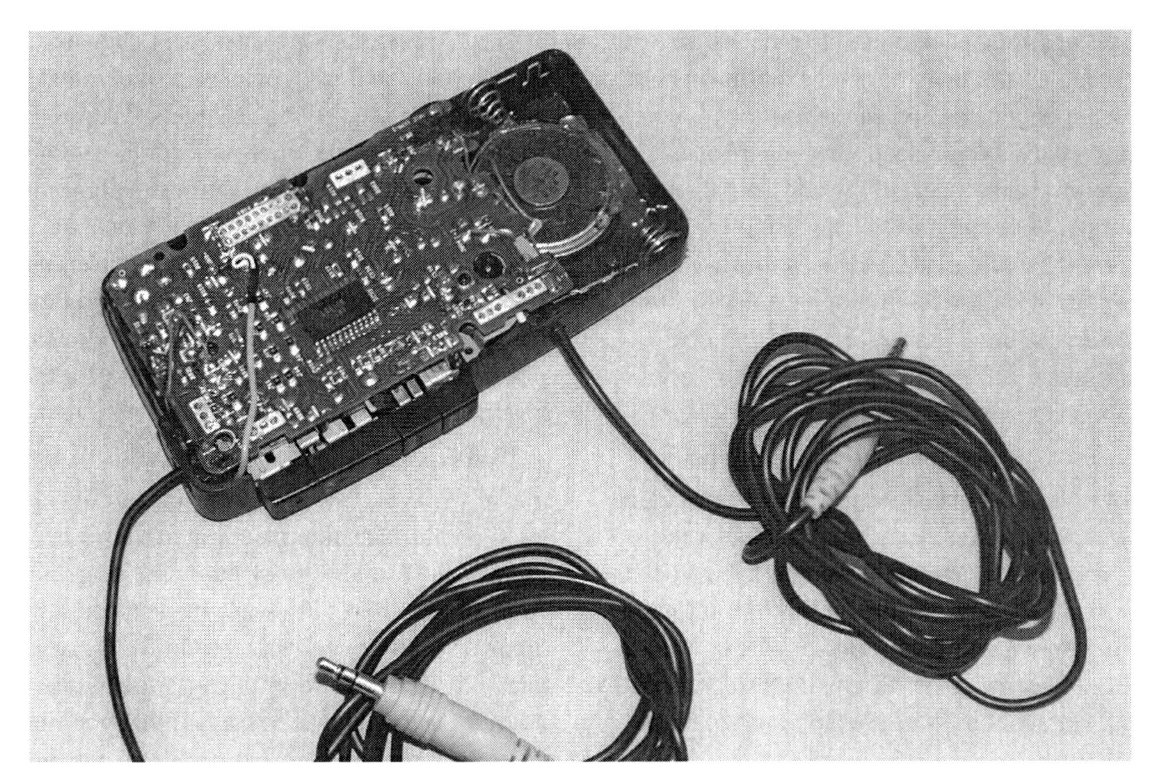

Figure 10-14 *Motor output wires added to the recorder.*

Figure 10-15 *The dream director ready for use.*

yourself up? Headphones are the obvious first choice and ensure that your odd experiments are not forced on anyone else in the room with you, but personally, I can't sleep with headphones stuck to my head, especially when sleeping on my side. If your audio device has a built-in speaker, you could place it directly under your pillow, but most devices do not have that feature, and then you run the risk of pressing buttons while you are asleep. A pillow speaker is very easy to make.

Pillow speakers can be purchased, but that's not my style, so I will show you how to make one that will work perfectly with this and any other audio experiment you might want to try. You can find a small speaker like the one shown in Figure 10-16 in most small radios, audio toys, or even a pair of old headphones. All you need to do is find a small container to fit it into and then solder a headphone jack to the speaker so that it can be plugged into the headphone jack on your audio player. I found a deodorant container that fit the speaker perfectly, so that is what I used.

The completed pillow speaker is shown in Figure 10-17 and also includes a switch that either can be used to shut off the speaker or can be fed back to some other project that requires user feedback. I did not have to make holes in the cabinet for the speaker because it was loud enough just as it was, and it does not take much sound at all at night when your head is right over the speaker. Another simple method for making a pillow speaker is to just take one side of a large headphone system and remove the strap part.

Now you are ready to dive into your favorite movie while you sleep or talk yourself into a lucid dream. The completed rig is shown in Figure 10-18, and it looks more like something from an anarchist's briefcase than something that belongs in a sleep laboratory! Oh well, just don't pack this device in your luggage when going on vacation, and you will be okay. What you plan to record is your business, but here are a few tips to get the best results without waking yourself up.

Figure 10-16 *Making a simple pillow speaker.*

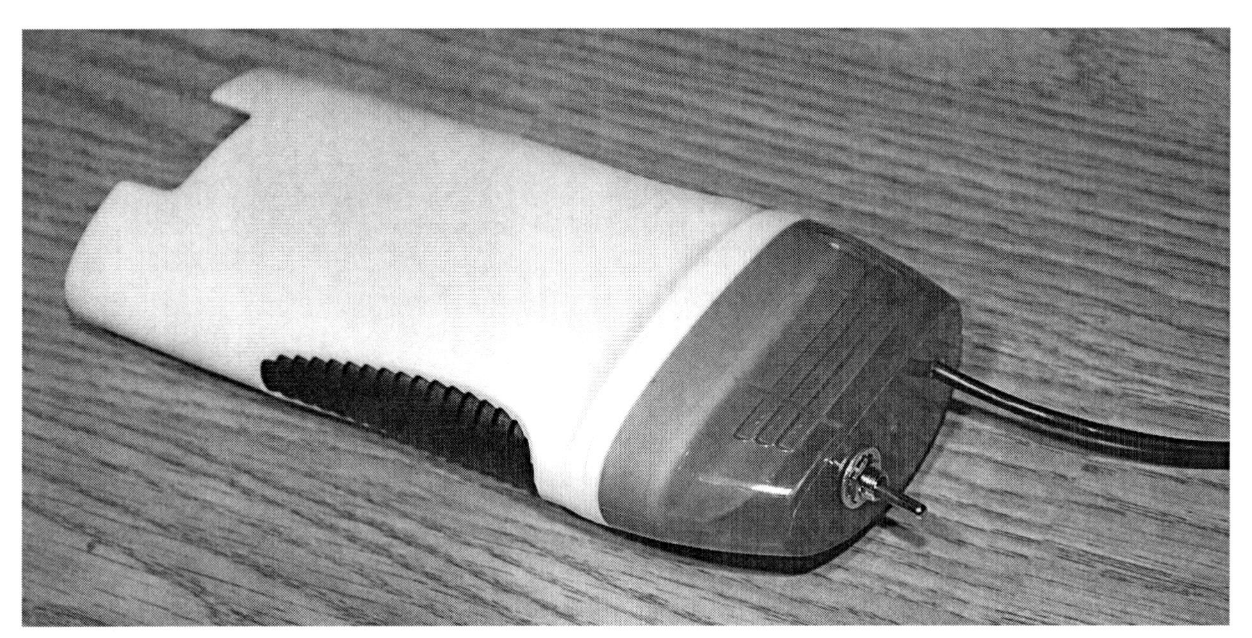

Figure 10-17 *The completed pillow speaker.*

If your audio starts suddenly, it will likely wake you up, especially if your audio file is some fast-paced action scene from your favorite movie. Practically any audio editing software can be used to create a *ramp*, or volume fade-in, over time so that your audio file starts at zero volume and increases gradually in level over a period of a minute or two. Figure 10-19 shows an old version

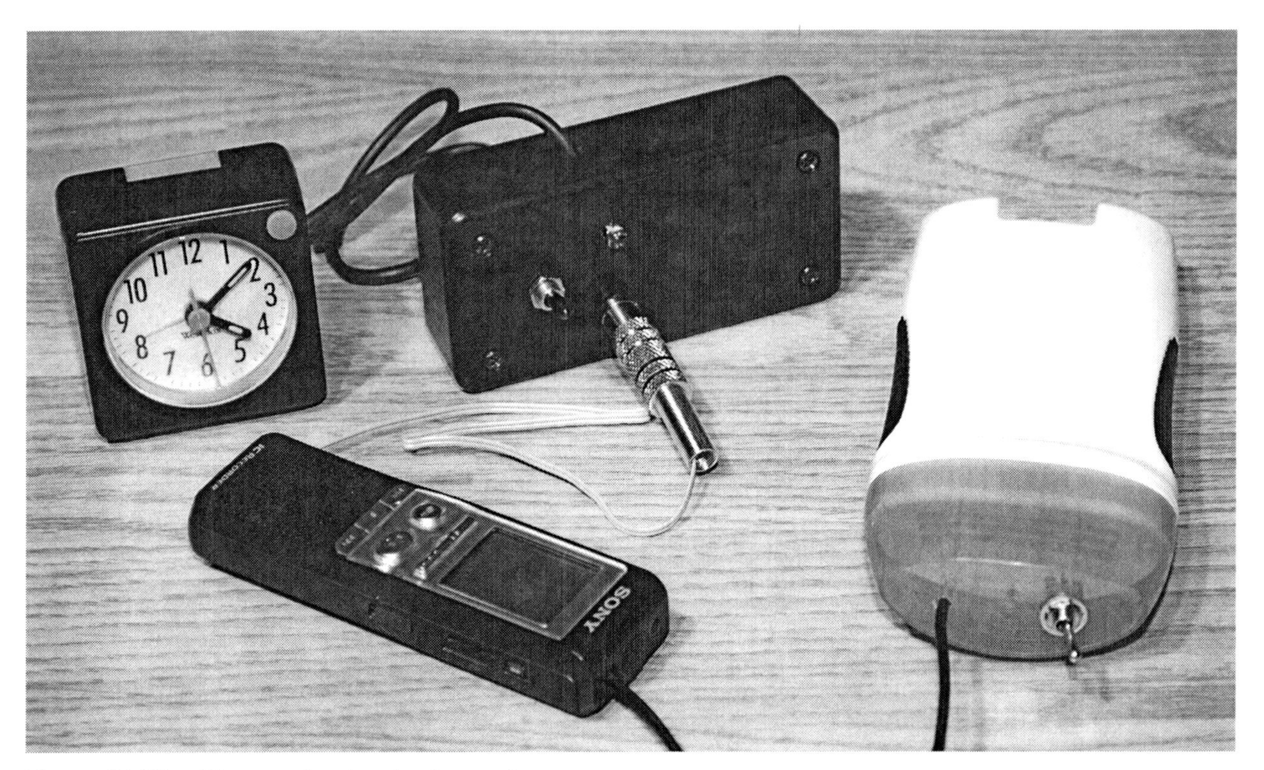

Figure 10-18 *The completed audio dream director.*

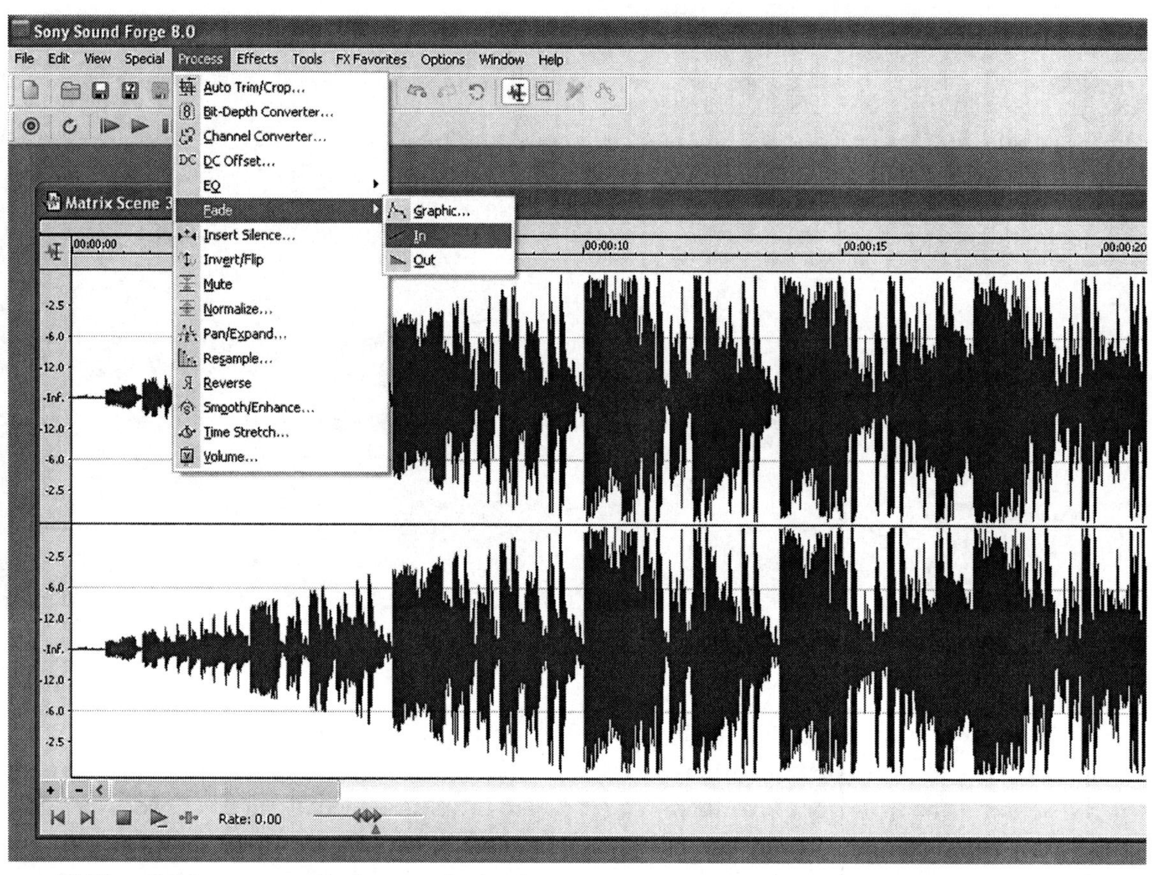

Figure 10-19 *Editing your audio for a gradual volume ramp.*

of Sony Sound Forge applying the "ramp in" filter to a recording of Neo fighting one of the Agents in *The Matrix*. Now the audio file comes in slowly so that it will not wake me from my sleep when the director triggers playback.

I hope that you have fun with this project; I know I did. The audio dream director can be a powerful tool for manipulating your dreamscape or even inducing a lucid dream if used properly. Try your favorite movie scenes or even overdub your own voice telling you to have a lucid dream like some kind of guru. How about sleep learning or self-help content? The possibilities are endless, and as long as you trigger the audio in your last stage of REM sleep and practice proper dream recall, you are guaranteed to see some results with this device. Of course, the dreaming mind is a strange animal, so don't be surprised it that *Matrix* fight scene induces a dream about you watching TV rather than being in the movie!

Project Eleven

Light-Sensing Lucid-Dream Mask

This project will detail the creation of what many consider to be the Holy Grail of lucid-dream hardware. By combining a circuit that monitors eyelid movements with a microcontroller, you end up with a fully self-contained lucid-dream mask or goggles that can detect your REM stage of sleep and send signals to you in the form of flashing lights or trigger any external device such as the audio system shown in the preceding section. By basing the trigger on actual rapid-eye movements, you are guaranteed to get your lucid-dream signals when you need them. These lucid-dream masks, or "dream goggles" as they are sometimes called, have been available for years and were made popular by Stephen LaBerge with his NovaDreamer product.

Our version of the lucid-dream mask will be presented using two different methods of detecting rapid-eye movements directly from the user's eyelids. The "classic" version will work much the same way that the commercial products do—using an analog amplifier fed by the output of an infrared transmitter and receiver pair to detect small changes in voltage as the dreamer's eyelids move back and forth. Another method, using a tiny accelerometer IC, actually will detect the eyelid movements directly, which makes the device so much easier to align and get working the first time. I have been experimenting with this alternate dream-mask design for several years now and find it to be more forgiving and accurate than the infrared REM-detection method, even when compared with some commercial units. Of course, you may have reasons to prefer experimenting with the classic system, so both versions will be fully detailed in this project.

In both versions of the dream mask, the output from either the infrared phototransistor or the accelerometer will be fed into a microprocessor for detection so that the signal can be further processed and handled by the user IO code. When 20 or more REM sleep movements have been detected from the sensor, the microprocessor will go into signal mode and begin flashing a visible LED 100 times so that the dreamer can learn to become aware while dreaming and enter a lucid dream. The microprocessor also aids in setup mode, allowing the user to fine-tune the eyelid sensor system by using instant feedback while wearing the mask or goggles. A false-trigger pushbutton is also added so that the unit can be reset if it goes off while he or she is still awake.

This project is presented in two parts because it does require a bit of adjustment and testing to get it set up and working perfectly. The REM-detection mask and related circuit will be built first and tested so that you can decide on which microcontroller to work with or even choose to feed the output directly into a computer for more control and ease of programming. The microcontroller part of this project will be presented next with included source code written in Basic for easy porting to any microcontroller or language. The completed unit will function as well as (or possibly better than) some of the commercially available dream masks, and it will cost you under $20 to build if you already have access to a microcontroller programmer. If you have never worked with microcontrollers such as AVR, PIC, or Basic Stamp, then don't worry; the hobby is inexpensive, easy to learn, and will be explained in more detail in the next part of this project.

Are you ready to become an *oneironaut*, one who travels in dreams? If so, then you will enjoy this project because I have found it to be the most effective way to train yourself to have a lucid dream.

The classic version of the REM-detection unit uses an invisible infrared LED (LED1) to shine light on the user's eyelid so that the reflection can be picked up by a matching phototransistor (PT1) and sent to an amplifier as a changing or modulated voltage. The schematic for this version is shown in Figure 11-1. Invisible light must be used because you don't want to have visible light shone into your eyes all night long. The output from the phototransistor (PT1) is fed into the input of IC1, the ML358 dual op amp. Op amps are great for designing high-gain amplifiers and filters, and because the LM358 contains two op amps, we will be using one as a filter to get rid of noise and the other to greatly amplify the voltage from the phototransistor. The filter is of the low-pass type, which means that a lot of noise induced by ac devices and rf interference will be eliminated so that there will be less noise in the amplified signal. The movement of your eyelids is very slow (1 to 4 Hz), whereas ac hum and other electrical noise will be at a much higher frequency (30 Hz or greater).

The variable resistor (VR1) is used to set the offset of the amplifier so that the signal can be adjusted to compensate for differences in many things, such as ambient light, placement of the

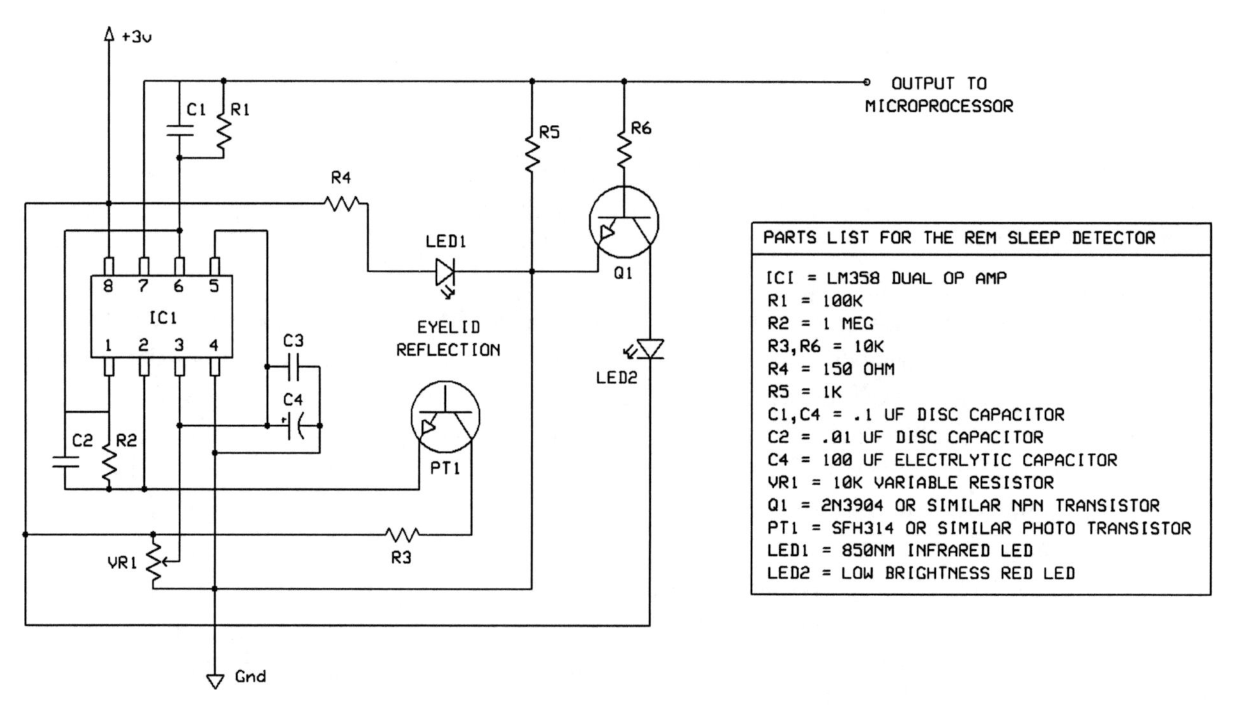

Figure 11-1 *REM-detection device schematic with phototransistor.*

mask on your face, and even the shape of various users' faces. The idea is to set up the system so that slight movements detected on the eyelids return a larger change in voltage so that the microprocessor or computer analog-to-digital converter has an easier time with the raw data. This adjustment of VR1 is critical to operation of the unit and can be quite quirky to get right, so a precision variable resistor is best used because it allows for a lot of fine-tuning.

To aid in the setup of the device, Q1, R6, and LED2 allow a visual indication of the feedback from the amplifier. When you have the detector set up perfectly, the visible LED will flicker slightly as you move your eyelids slightly, an indication of a good change in voltage from the amplifier and filter. Once you are happy with the operation of the unit, Q1, R6, and LED2 can be omitted from the final design because the microprocessor code also will contain a test routine to help you set up the device in real use.

The entire system is designed to run on 3 V dc, so you can power it for several nights on a pair of rechargeable AA or AAA batteries. Also, dc operation keeps unwanted noise from the circuit and isolates your melon from the ac power lines at night. This is good because an electrical storm while you are sleeping could be bad news if your dwelling has a direct strike. I highly recommend that you first build this system on a solderless breadboard because it can be finicky to set up the first time, especially if you decide to use one of the many other op amps available. The LM358 is just a generic op amp that can run from a single voltage supply, but there are many available that would work just fine and possibly give better results. Experimentation is always the goal of this book, so don't be afraid to wander into your own territory and try something different.

The infrared LED and phototransistor pair becomes the heart of this unit, and there are many options available that will work perfectly once you set up the proper distance and reflections spot. The only requirement is that the infrared LED and the phototransistor work at the same wavelength, which will be between 850 and 1050 nm for infrared light. Infrared LEDs from old TV remote controls will work perfectly, and practically any infrared phototransistor will have the correct bandwidth, so it should not be hard to find a pair that work together. RadioShack and many electronics hobby stores will sell them in pairs, and they will look much like the pair shown at the top right of Figure 11-2. The three black-boxed units across the center of the figure are called *position detectors* and contain both the infrared LED and the matching phototransistor in one enclosure. The position detectors work well and are a bit easier to align because they are already set up at the correct angles, but be aware that some of these have a digital output and may not work well in this design. The large unit to the left of the figure has the part number QRB1113 or QRB1114 and was found to work quite well in this design. The unit in the center right of the figure is designed so that an object placed between the two walls will break the beam, and this unit can be hacked apart to reveal the small matched pair, as shown in the bottom of the figure. These beam-breaker units are very common in printers and photocopiers. By using trial and error, you should have no problem selecting the best matched pair from your junk box.

The breadboarded circuit is shown in Figure 11-3, where I use my finger to test the multitude of infrared phototransistor and LED pairs I found in my junk box. The matched pair in this figure was taken from an ear clip heart rate monitor sensor I found on an old exercise bike computer, and it worked very well. Also notice the precision trimpot variable resistor used for VR1. These precision units allow for a very fine control over the entire scale, and they make setup of the amplifier much easier. Another thing to note is that ambient light will have a huge effect on the workings of this circuit, so it is best to set it up in a dark room. Sunlight and many room lighting systems also will contain infrared light, so the phototransistor could act much differently in a

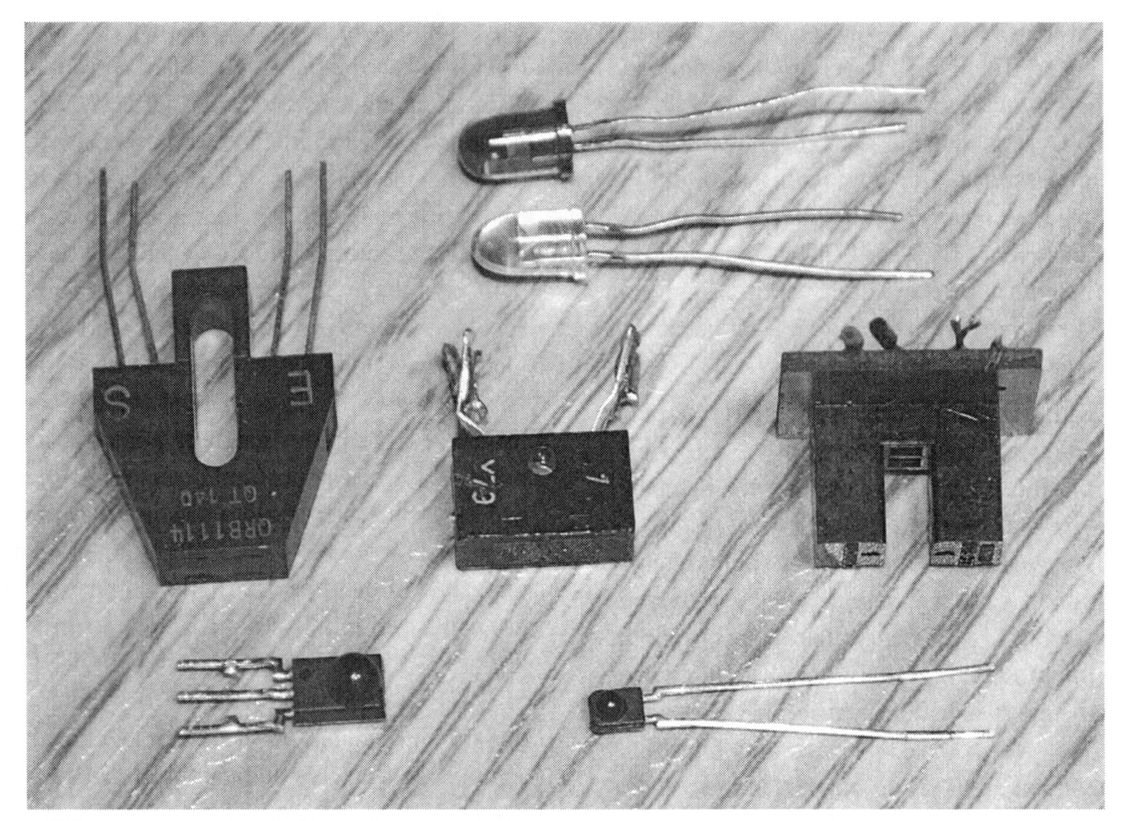

Figure 11-2 *Trying various infrared detectors.*

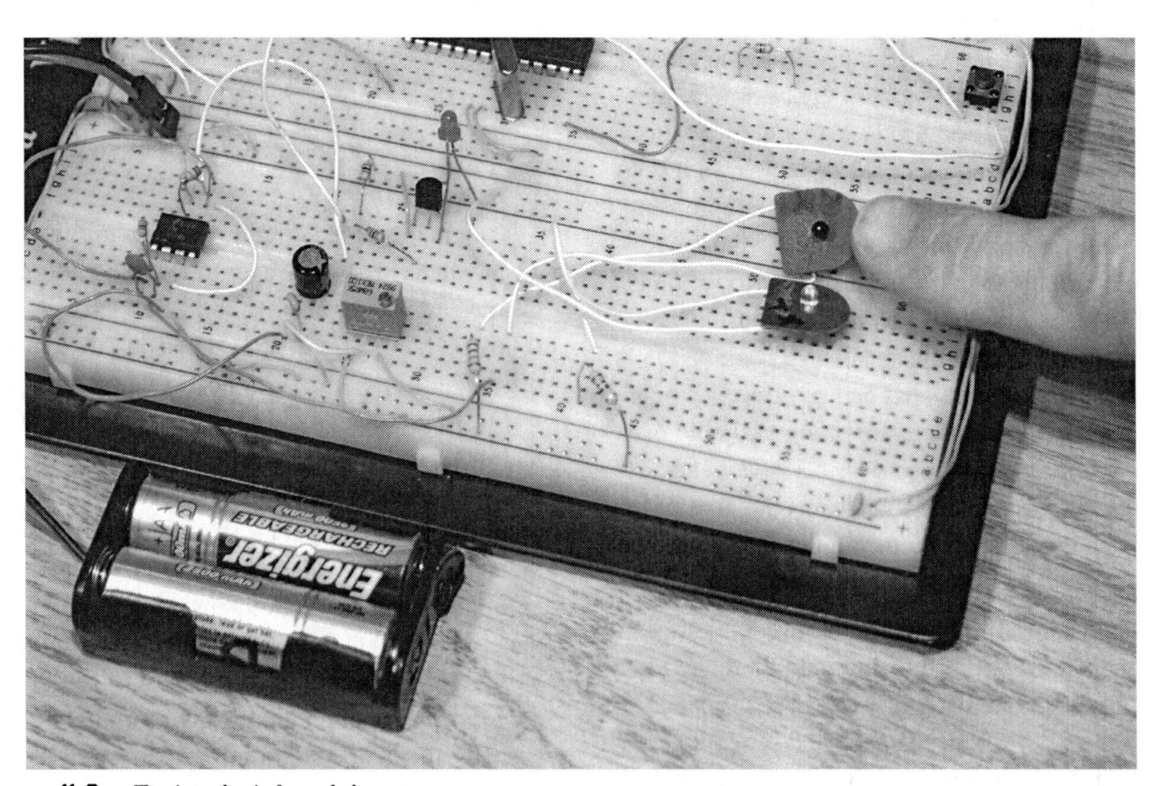

Figure 11-3 *Testing the infrared-detector response.*

lighted room, requiring more setup. Try to point the phototransistor and LED at an angle so that they meet at a point that is about 1 in away. This will be approximately the distance from your eyelid to the matched pair once the unit is built into some type of goggle or mask housing.

Most of these do-it-yourself (DIY) dream masks are built into small swim goggles like those shown in the lower part of Figure 11-4, but I found a ski mask to be so much more comfortable when I was wearing it for extended periods. You want the mounting system to be as loose as possible on your face yet secure enough that it does not shift as you move around in bed. Swim goggles are certainly secure, but the strap seems to be too tight, and the rubber eyepieces are not really designed for comfort as much as they are for creating a watertight seal. If you replace the rubber head strap with some lighter elastic and drill holes in the goggle casings, they are not too bad, but the ski mask is a much better solution and will feel a lot more comfortable and secure. Using a ski mask also gives you a bit of room to mount the battery pack and electronics so that you can make a self-contained unit rather than having wires in your bed.

If you are a junk collector like I am, then you probably have a bunch of infrared beam-breaker-type sensors in your junk pile like the one shown in Figure 11-5. These small units can be broken apart to release the phototransistor and matching LED by carefully prying open the case using a small knife. Before you start hacking, look at the case so that you can identify each component, which will have either an arrow pointing from the LED to the phototransistor or markings that read "AK" (anode, cathode) and "CE" (collector, emitter). If there are no markings at all on the shell, you can just guess by trying each component one way or the other until you get some response. This will not fry the parts because the voltage and current are very low in the circuit.

Figure 11-4 *Selecting a comfortable mounting system.*

Figure 11-5 *Hacking a photo sensor into its matched pair.*

Figure 11-6 *Positioning the photo-sensor components.*

I decided to use the tiny phototransistor and LED that I hacked from the sensor shown in Figure 11-5 because they worked very well. There were no markings on the case, so it took a while to figure out by trial and error that the darker unit was the receiver and the lighter unit was the transmitter. Figure 11-6 shows the matched pair mounted to a bottle cap using some double-sided tape to hold them in place. I also had to cut a bit of the black plastic case out and place it between the two components to keep the beam from reflecting directly from one to the other rather than bouncing off my eyelid first. The small sensor shown on the right in the figure is one of the all-in-one position sensors and is basically the same thing as my bottle-cap hack but ready to be used. You can find these all-in-one units at any electronics supplier by searching for optical position sensors or distance sensors.

After testing various optical-sensor configurations and deciding that the bottle-cap hack was the best one, I placed the components on a small perf board so that the system could be connected to my ski mask for further testing. Figure 11-7 shows the small perf board with the programmed microcontroller (discussed in the last section of this chapter) ready for operation. It is a good idea to get the analog part of this project working before you move on to the microcontroller or computer interfacing because this is the area that will take the most effort to get working properly. I tested my ski mask with the optical sensor while the circuit was still on the breadboard before getting to this stage in the design. If you have an oscilloscope handy, feed the infrared LED with a small 1.5-V watch battery, and then plug the output from the phototransistor into your oscilloscope so that you can see the analog results directly while you wear your face mask and move your eyes back and forth (one closed eye). This will give you a direct reading of the responsiveness of your optical sensor's output. Some careful alignment might be necessary depending on the beam width and field of view of each of the components.

The proximity and angle of the phototransistor and infrared LED are very important for optimal response and may take quite a bit of fine-tuning to get working properly. If you are using a self-contained photo sensor, then only the distance from the edge to your eyelid will be important, but if you are using a separate phototransistor and LED, then both the distance and angle will matter. Figure 11-8 shows my hacked bottle-cap sensor placed inside the ski mask lens at the approximate position where my eye will be. I put on the mask and drew a mark on the lens while looking in the mirror so that I could find this position. A line was drawn around the cap so that the lens could be cut out using a routing bit on a Dremel tool. I could have just glued the cap

Figure 11-7 *Placing the circuit on a perf board.*

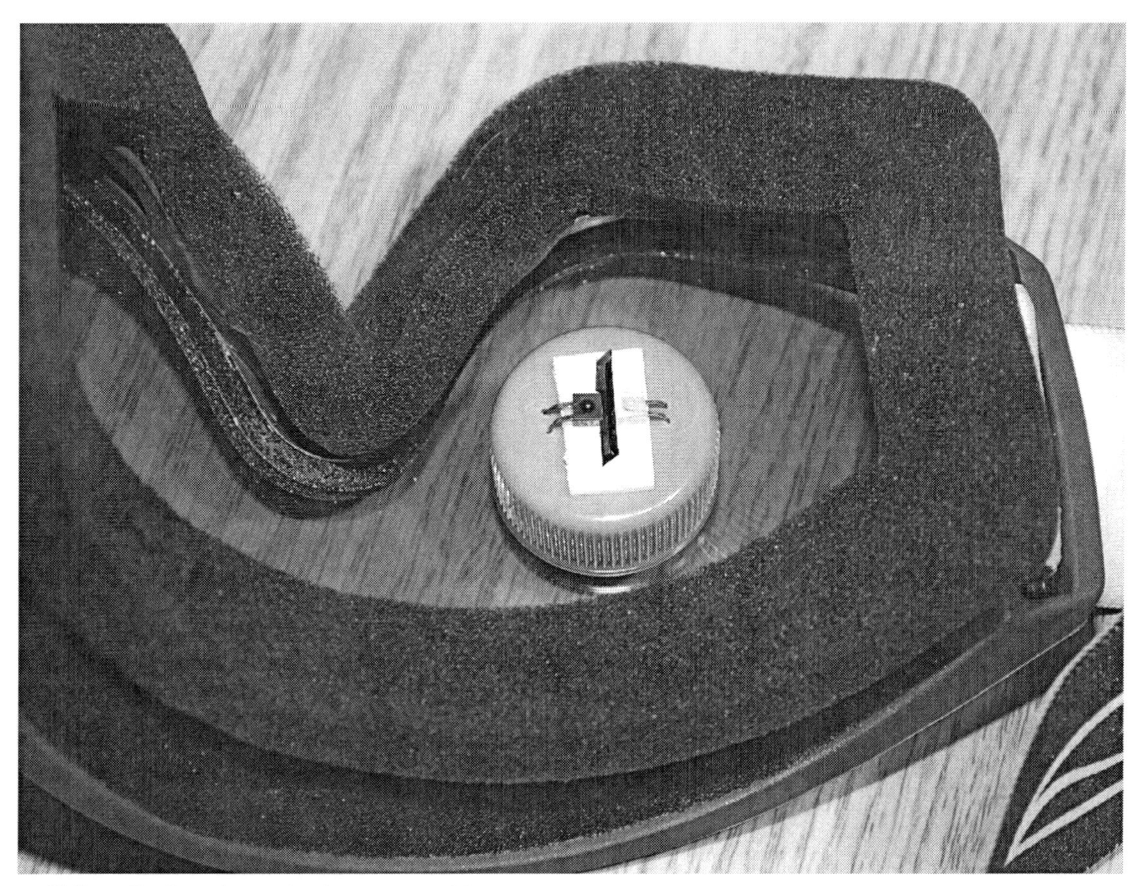

Figure 11-8 *Finding the optimal sensor position.*

directly to the inside of the lens, but it was a bit tall, and by cutting a hole in the lens, I had some control over the angle if I decided it needed to be adjusted later.

With the bottle-cap sensor installed in a hole through the ski mask lens (Figure 11-9), it was easy to set the distance as well as the angle so that an optimal response could be received from the infrared light reflecting off the user's eyelid. Again, this was done in a dark room while looking at the indicator LED on the breadboard circuit and moving my eyes back and forth until the LED gave the greatest response. If you find it annoying to look through one eye in the dark and tweak the variable resistor, then find a human helper to wear the mask as you fine-tune the system. Most likely the adjustment will work for all your test subjects because all human faces have the same basic geometry. Once you have the system mounted to your goggles or mask and working properly, you can go ahead and mount your components on a more permanent home for installation onto the mask or into some small box.

The completed optical-based REM-detection system is shown mounted to the ski mask in Figure 11-10. I split the batteries up to balance the weight on the mask and placed the circuit board in the center for easy access to the reset button in case of false triggering. The mask certainly was comfortable and worked well in all positions except for face down, of course. I decided to keep the lens clear rather than paint it black so that I could get up at night and move around without having to remove the mask and then go through the microcontroller test procedure each time. Adding the circuit board into a box would have made the unit more robust, but since a project is never complete, I decided to leave the board out for easy access when it came time to add or modify the circuit. Also notice the visible LED on the side opposite the sensor unit. Again, it's placed directly in front of

Figure 11-9 *Inserting the cap through the lens hole.*

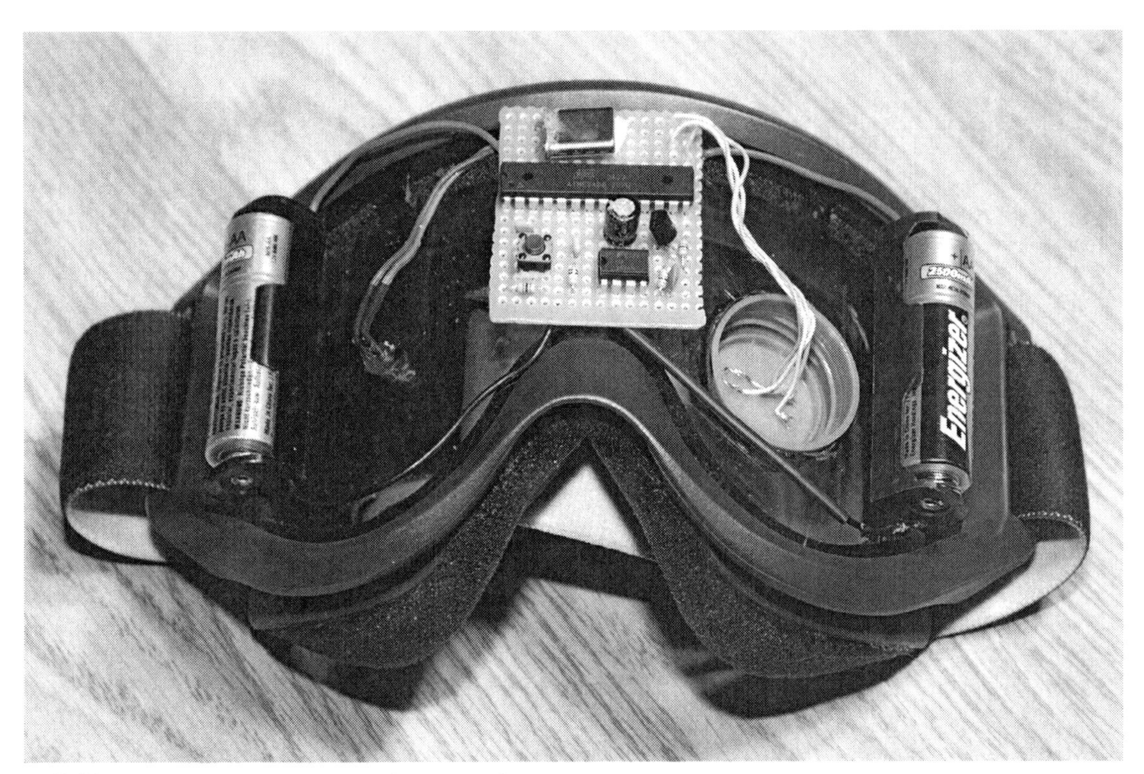

Figure 11-10 *The completed infrared dream mask.*

my eye for optimal brightness. The indicator LED will be discussed in the last section of this chapter when dealing with the microcontroller and programming. You also might notice that there is no on/off switch on my dream mask. This is so because I usually recharge the batteries after each use and just let the unit run all night, which is usually how I use the system. If your plan is to wake an hour before your normal wakeup time and then use the mask, a power switch would be handy.

Once your dream mask has passed all your tests, it's time to become an oneironaut and make the journey to your other world. As you can see in Figure 11-11, the lucid-dream mask is more than just a powerful tool; it is also a cool Halloween prop or "steam punk" accessory! Besides alignment and initial testing (discussed in the last section of this chapter), the only other concern is sleeping with your mask strapped to your face. I found the ski mask comfortable, and since I normally sleep on my back, wearing it to bed was not a problem. Where I initially had some difficulties was being too excited to fall asleep easily owing to the possibility of an easy lucid-dream journey. It actually took a few nights of going to bed wearing the mask to end up with a good night's sleep because I ended up removing it to actually fall asleep. On the first full night with the mask, I woke right away when the LED started flashing—again too eager to see results. I ended up enhancing the microcontroller code to hold back detection until 6 hours after starting the unit (discussed later) so that I would not be awakened from an early REM stage. After several attempts and adjustments, I found that the mask does indeed fulfill its purpose as long as you can remember to "see the signs" in the dream world. These issues will be discussed at the end of Section Two when you complete the microcontroller part of this project.

Now let's look at an alternative to the infrared light-sensing system that uses an analog accelerometer IC to detect actual eyelid movements.

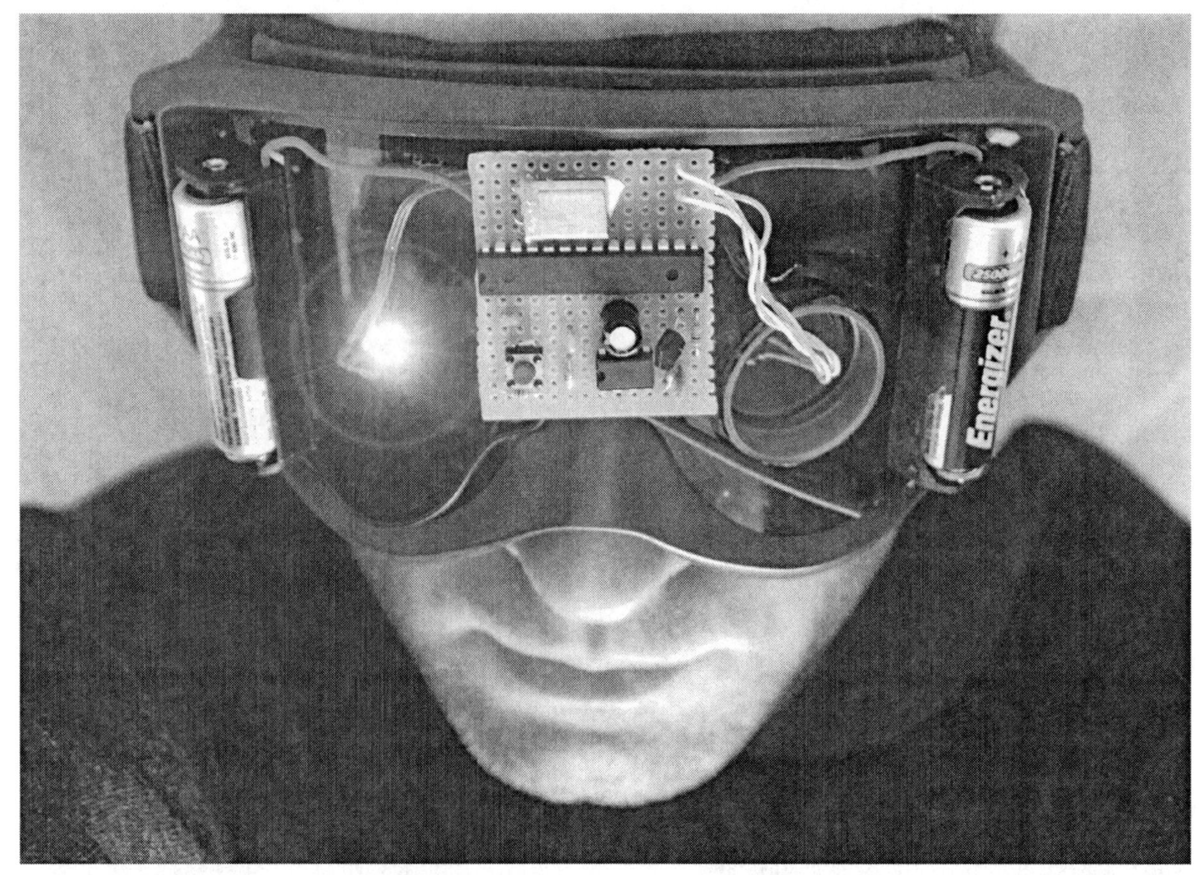

Figure 11-11 *Another oneironaut ready for takeoff!*

Project Twelve

Motion-Sensing Lucid-Dream Mask

This version of the lucid-dream mask project will replace the infrared light sensor with an actual motion sensor that can detect the smallest change in velocity. I came up with this system after finding the infrared version to be a bit quirky when it came to getting proper alignment. Even the commercially available units sometimes can be finicky when it comes to moving around in bed or having ambient light enter the phototransistor, so I wanted to try an alternative method of detecting eyelid movements. By using an inexpensive tin IC called an *accelerometer*, I was able to place the unit directly on my eyelid and feed its output directly into the microprocessor, bypassing all analog circuitry and making the unit completely impervious to ambient light and all electrical noise. It sounds uncomfortable to have something in contact with your eyelid, but the sensor is so small and lightweight that it feels no different from how the cloth on the sleep mask originally felt, so it is not an issue. The good news is that this system is far easier to set up and get working properly because the only data sent to the microcontroller are actual eyelid movements, and changes in the subject's position do not cause the system to fail owing to alignment changes.

An accelerometer is a tiny IC that can detect the slightest change in direction, and accelerometers are used in everything from self-balancing scooters (e.g., Segway) to impact sensors in vehicles to activate airbags. These amazing devices are so sensitive that I was able to connect one to a microcontroller and actually digitize my voice just by talking near the unit as it dangled from a wire on my breadboard! Yes, it actually converted the vibrations from the sound into digital data in real time. So you can imagine that detecting the movement on your eyelid would be a very easy chore for such a sensitive device. These little accelerometers are also very inexpensive now and can be purchased from many electronics distributors.

Figure 12-1 shows the accelerometer (the tiny block on the left) I decided to use for this project, the ADXL 202E, and my analog device. This $10 part actually has two accelerometers to detect *x*- and *y*-axis movement, but I will be using one channel because it was all that was needed. You can even find three-axis accelerometers for the same price, and if you get creative with your microcontroller code, you could filter out side-to-side movements, making your eyelid detector extremely robust to any stray head movements. The strip of even smaller accelerometers on the right in the figure holds three-axis units I plan to experiment with later as I get more complex with my microcontroller coding. I even have found

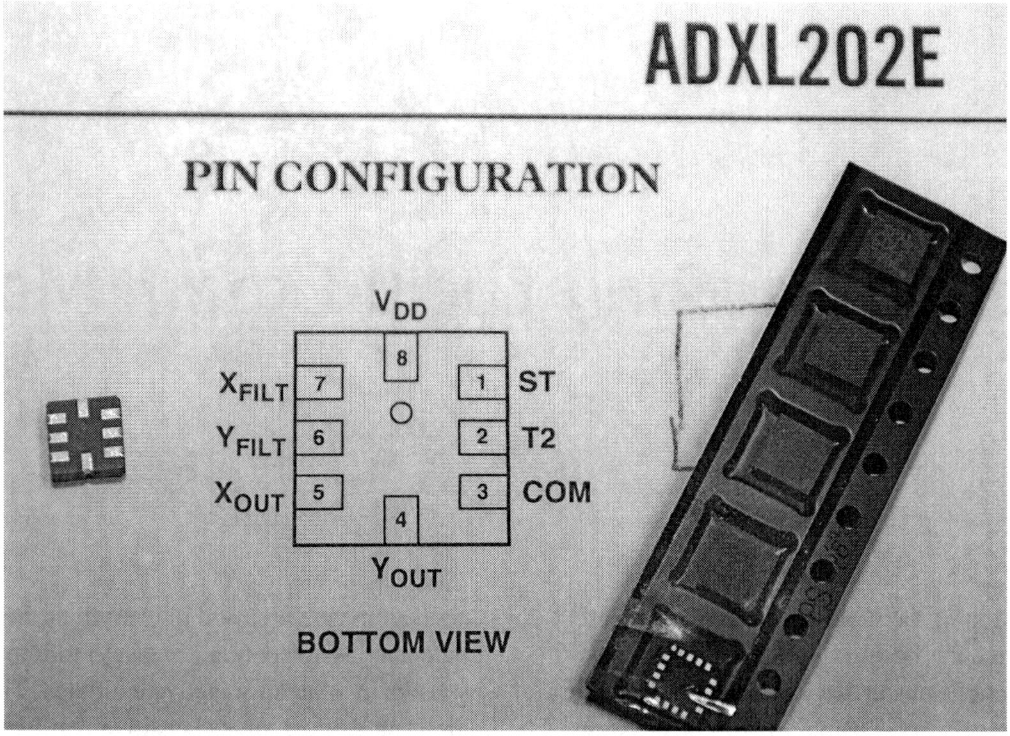

Figure 12-1 *A very tiny and sensitive accelerometer IC.*

these ICs to be sensitive enough to detect my pulse when placed in the optimal spot on my neck!

It really does not matter which brand or style of accelerometer you choose as long as it has a pulse-width-modulation (PWM) output so that it can be measured by your microcontroller as a digital signal. Pulse-width modulation is very simple: A series of pulses is sent out of the device at a constant rate, but the duration (duty cycle) of the high pulse is proportional to the detected motion of the device. For instance, if no motion is detected, the pulse may stay high for 1 ms and then drop back low for another millisecond. This would mean that pulses are sent every 2 ms (while they are using my device). Now, if motion is detected, the pulse may stay high for only half a millisecond and then drop back low after 1.5 ms, indicating that a small change in motion has been detected. Notice that the total pulse time is still exactly 2 ms, but the duty cycle or high time has been changed.

This PWM scheme is very easy to measure in a microcontroller because you simply wait for the rising edge of the pulse and then start a running counter until the output drops back down to zero. It becomes even easier in my system because I don't care about the actual value of the reading, just that it has changed from the last reading, so the code is very simple. Operation of the microcontroller code will be explained in the next section, so let's just figure out how to get the accelerometer set up to detect eyelid movements for now.

Although the ADXL is larger than many of the accelerometers available, it still seems dwarfed by the other semiconductors on my breadboard (Figure 12-2). I did not include a schematic because there really is not much to show; the accelerometer basically feeds its output directly to the microcontroller. The two resistors and the capacitor connected to the accelerometer are used to set the filter and response time to 2 ms as per the datasheet. Your accelerometer likely will have different methods for setup, but not to worry,

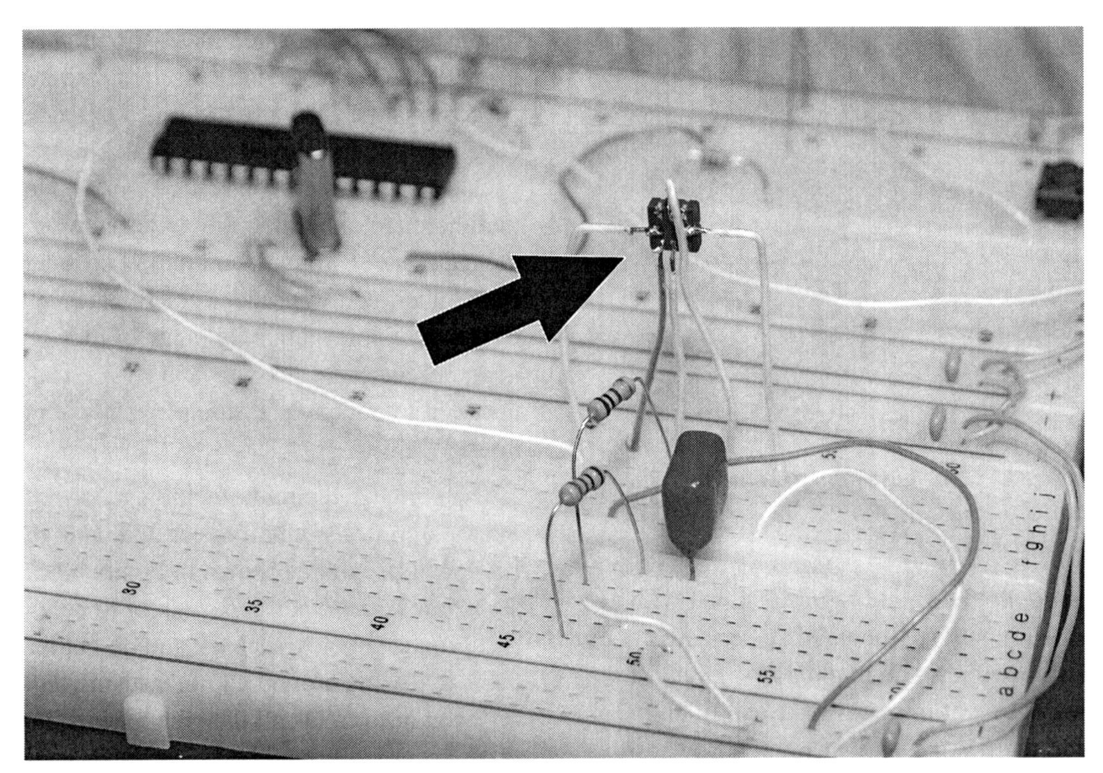

Figure 12-2 *Testing the accelerometer on a breadboard.*

because the microcontroller code is very forgiving; all it does is compare the PWM signal coming out of the accelerometer. Much faster or slower sampling rates will work just fine because it is change, not time, we are comparing.

It is best to start on a breadboard and include the programmed microcontroller so that you can make sure you have soldered the wires correctly on the tiny accelerometer package. If you have yet to work on the microcontroller part of this project, then you can just feed the output into an oscilloscope and watch the waveform change as you tap on the body of the accelerometer. Another trick I use to check if these things are functioning is to connect a piezo buzzer directly to the output pin and listen to the digital nose change as I tap on the device. An ugly hack, I know, but it does work! Once you have determined that there is indeed a signal spewing from the accelerometer, you can proceed to mount it on a small perf board along with the few needed semiconductors that make it function. The goal is to have only three wires coming from the accelerometer unit: power, ground, and output signal.

To keep the accelerometer unit as light as possible, I decided to cut a small bit of perf board to fit into a plastic bottle cap, as shown in Figure 12-3. In this way, I could add a bit of felt to the cap and just rest it on my eyelid for movement detection. This worked very well and did not even require a sleep mask when sleeping on my back because the cap just stayed there. There are many ways you could mount the accelerometer, including just a bit of heat shrink with the thin wires coming from the device. The wires should be thin and flexible as well because you do not want to hinder the movement of the accelerometer unit as your eyes begin to move back and forth. The other bottle cap shown in the figure will be used to carry the visible LED so that my sleep-mask installation has basically the same thing on both sides.

It took a bit of fine work to get the tiny accelerometer soldered down to the small

Figure 12-3 *Cutting a small perf board to fit the bottle cap.*

breadboard, as shown in Figure 12-4, but with a little patience, surface-mounted devices are not as bad to work with as you might think. Even after installing the two timing resistors and filter capacitor, there was plenty of room left over in the bottle cap. Maybe if you are a real wiz with a soldering gun, you could even include a surface-mounted microcontroller and 3-V button battery to make the entire device self-contained in a space no larger than a bottle cap! I opted for the easy way out and stayed with a dual inline package (DIP) microcontroller for now. Notice that there are only three tiny wires coming from the cap: power, ground, and output signal. Remember, you want the accelerometer to be as light as possible so that its range of movement is not restricted in any way and so that it does not need to put any pressure on your eyelid when in use.

The completed circuit, including the microcontroller, accelerometer unit, and visible LED, can be built as shown in Figure 12-5, ready for a new home in a sleep mask or to be used as is. If you sleep on your back and do not have a habit of tossing around all night, then you may not even need to install the components on a mask because the small accelerometer and LED units can just be placed directly over your eyelids if you place the circuit board remotely in a small

Figure 12-4 *Accelerometer and components in a bottle cap.*

Figure 12-5 *Accelerometer, microcontroller, and visible LED.*

box. Since this version of the REM-detection system is completely oblivious to ambient light, you do not have to shield the sensor or worry about nearby electrical devices that might induce noise into the circuit. I actually prefer using the device remotely, with the two caps resting on my eyelids, but I will show how a cloth sleeping mask also can be used for those who may sleep sideways or move around a bit at night.

Even if you decide to use the device without a mask or a face strap, you will need to add the microcontroller, batteries, and test switch in some type of case, as shown in Figure 12-6. Besides the microcontroller and LED-driver transistor, there is not much else to the control board, so it should fit in a small box along with the two AA or AAA batteries, an on/off switch, and the test/reset pushbutton. Since operation of the unit only requires turning it on and then pressing the test button to ensure that the sensor is working, you will never have to look at the box, so it can be placed remotely so that it is easily accessible while you are relaxed and "wearing" the accelerometer and LED units.

If your battery pack and black box are small enough, you can mount the system directly to a cloth face mask to create a self-contained unit, as shown in Figure 12-7. The cap with the accelerometer is on the right side of the figure, and the visible LED is on the left. Notice that I am using the accelerometer cap top down and the LED cap open-end down. This makes the visible LED brighter and allows the accelerometer to move easily as my eyes are moving. I later added some felt pads to both caps, which helped to keep them in place when I was sleeping on my side using the mask setup. As with all home-built projects, a little trial and error (hacking) probably will be needed on your part to find out what works best for you.

The control box is shown on the front of the cloth sleeping mask in Figure 12-8, ready to be used for some serious lucid-dream training missions. I found this version of the REM-

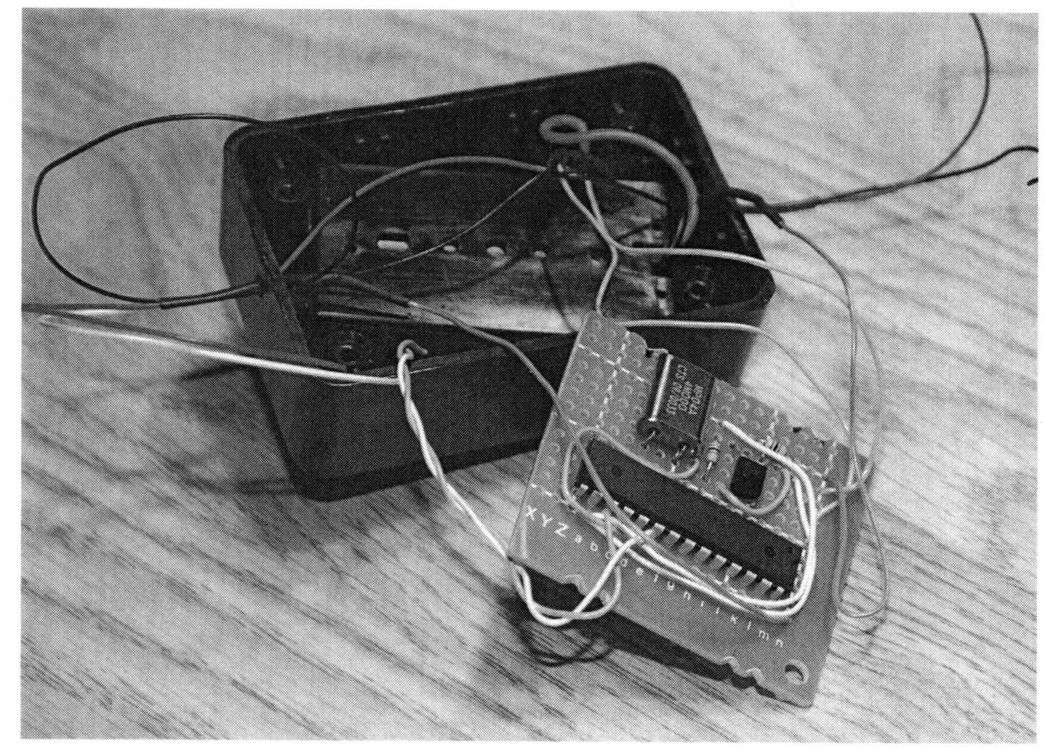

Figure 12-6 *Installing the microprocessor unit in a box.*

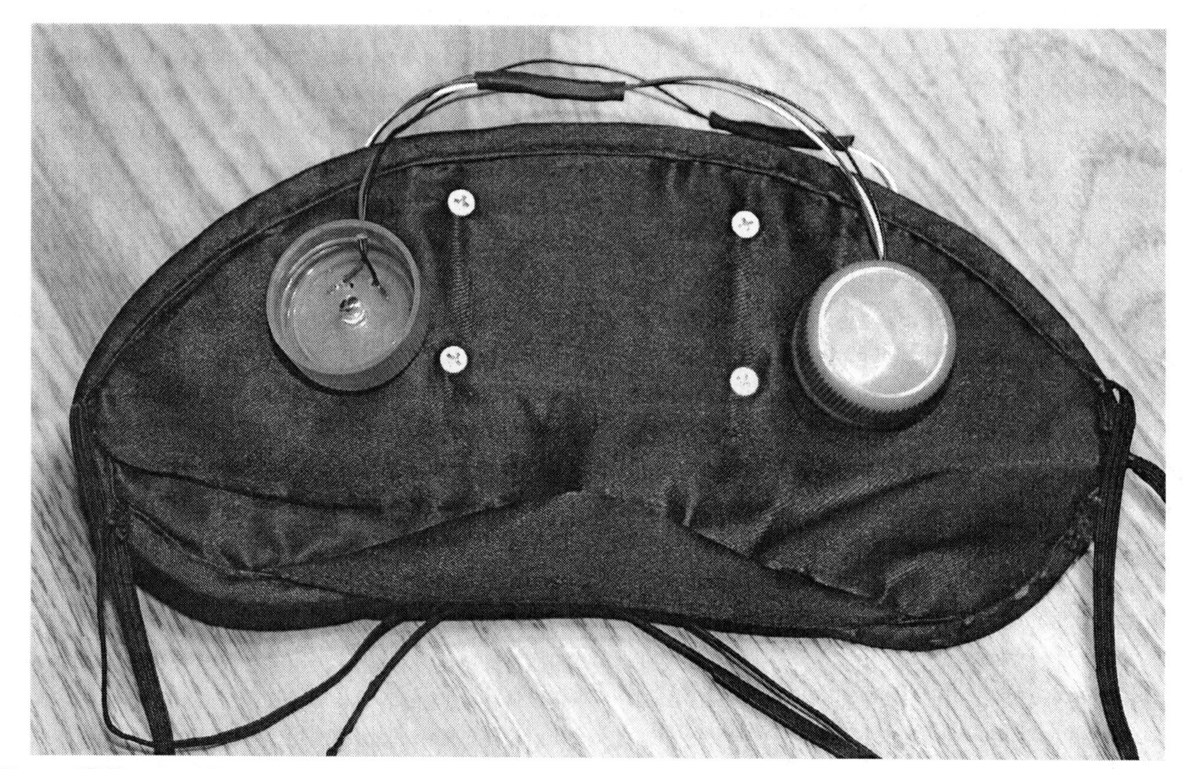

Figure 12-7 *Installing the hardware onto the sleep mask.*

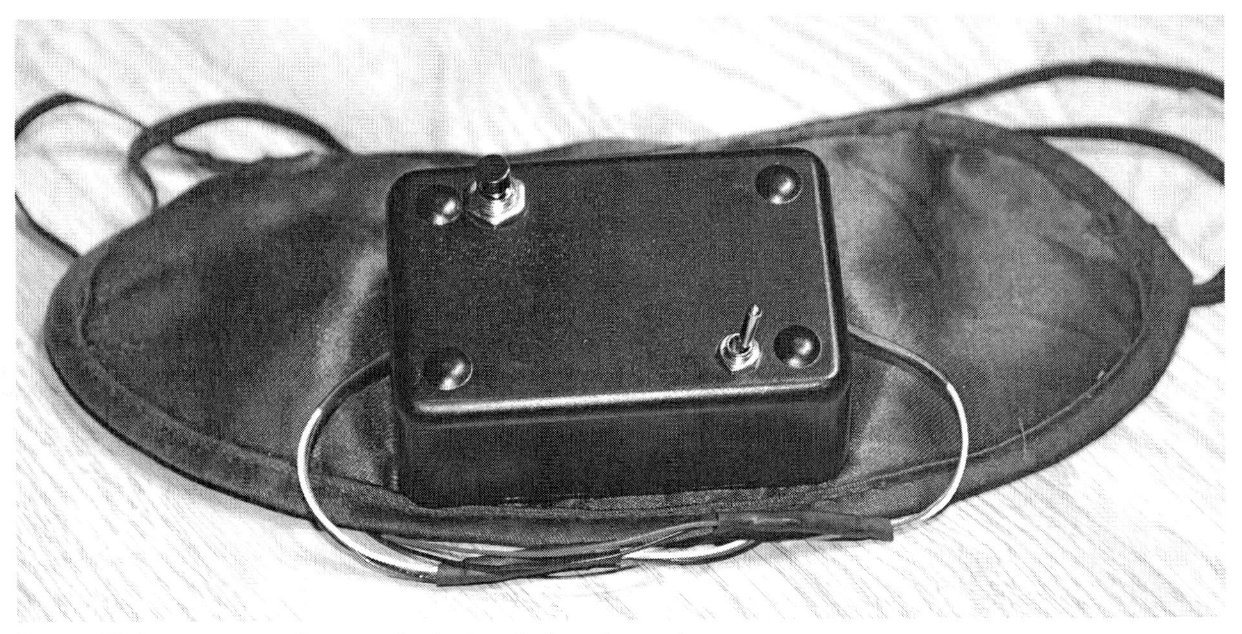

Figure 12-8 *The controller can also be installed on the mask.*

detection system to be much more robust and easier to set up than the infrared version presented earlier, and it has been a great tool for further exploring the dreaming world. No matter which unit you decide to build (how about merging both of them together), you will need to move on to the next step and program the small microcontroller to sort out the raw data from the sensor and control the visible flashing LED. If you have never considered adding microcontrollers to your arsenal of electronics know-how, then trust me, it is worth it and will open up doors you never thought possible. Next we will build the lucid-dream mask controller.

Project Thirteen

Lucid-Dream-Mask Controller

Now that you have a working REM-detection system based on either the photo sensor or the accelerometer, you will need a way to create a response from the data received. Using some fancy analog circuitry and a few digital counters, you probably could rig up a system that counted pulses from the detected eyelid movements and then triggered a flashing LED, but this would become a very quirky system with a large number of components. The lucid-dream-mask controller needs to perform a number of different operations to make your lucid-dream mask functional and easy to use.

First, the controller needs to either detect the difference in the analog signal from the phototransistor, or it needs to measure the change in PWM from the accelerometer. This job would require a lot of finicky analog circuitry to get working properly, but any small 8-bit microcontroller can do this job and not even break a sweat. After the detection phase, the controller must begin to count changes in order to avoid triggering the visible LED each time the user moves his or her eyes. A few movements may not indicate REM sleep, so the controller needs to look for continuous eye movements, which is another simple task for any microcontroller. In my version, I just count 20 eye movements. Once the controller has decided that the user is in REM sleep, the visible LED must be flashed a number of times (100 in my case) so that the user can look for this signal in his or her dreams. Often the signal will be something related to flashing light in the dream, so it takes a bit of practice and conditioning to recognize the signal.

Another task for the controller is to create some type of test procedure so that the user can verify operation of the unit and to reset the counter if a false trigger happens owing to accidental waking eye movements or simply because of waking up from REM sleep. In the version of the Basic code presented here, the controller does both functions from a single pushbutton so that it is very easy to use the system in the dark. If a false detection happens, the user simply presses the button, and the controller resets the system to start over. If the user holds the button down for 5 seconds, the controller puts the system into test mode and will flash the visible LED instantly every time there has been sensor detection. In this way, the user can verify that the system is indeed responding to eye movements. To exit test mode, the user holds the button down for another 5 seconds, and the system once again resets and starts working in REM-detection mode.

The source code is extremely simple and was written in Basic so that it can be ported easily to any other microcontroller or any other language. Before we dig into the source code, let's look at how simple the schematic for the lucid-dream-mask controller really is.

Although the schematic shown in Figure 13-1 is amazingly simple, the unit has a great deal of functionality, even with the extremely simple basic program that lives inside the chip. The Atmega88 from Atmel is an inexpensive 8-bit microcontroller that offers 23 IO pins, 8K flushable program memory, 1K SRAM, 20-MHz operation, a bunch of onboard peripherals such as an analog-to-digital converter and serial port, and a cluster of hardware timers. Of course, this is pretty standard for many of the low-end, low-cost 8-bit microcontrollers offered by various other companies such as Microchip as well. Because the controller program is kept to a minimum and written in Basic, you could adapt easily it to just about any low-end 8-bit microcontroller or even a Basic Stamp module from Parallax, Inc.

If you have never heard of an 8-bit microcontroller before reading this book, then think of it as a blank IC that costs you only five bucks and can be programmed to do anything imaginable from emulating a simple logic circuit to playing video games on your TV. No kidding! I have made single microcontrollers play music, record sound, control huge robots, log computer keystrokes, and even emulate entire retro 8-bit computer systems. Since the magic happens in your code, it will feel like the entire world is under your control once you feel the magic of the microcontroller. There are thousands of various microcontrollers available, some of them having as few as 8 pins and a few kilobytes of program storage, whereas others have hundreds of pins and megabytes of onboard memory. Our little Atmega88 is considered a smaller microcontroller, but it is still overkill for the simple program presented here.

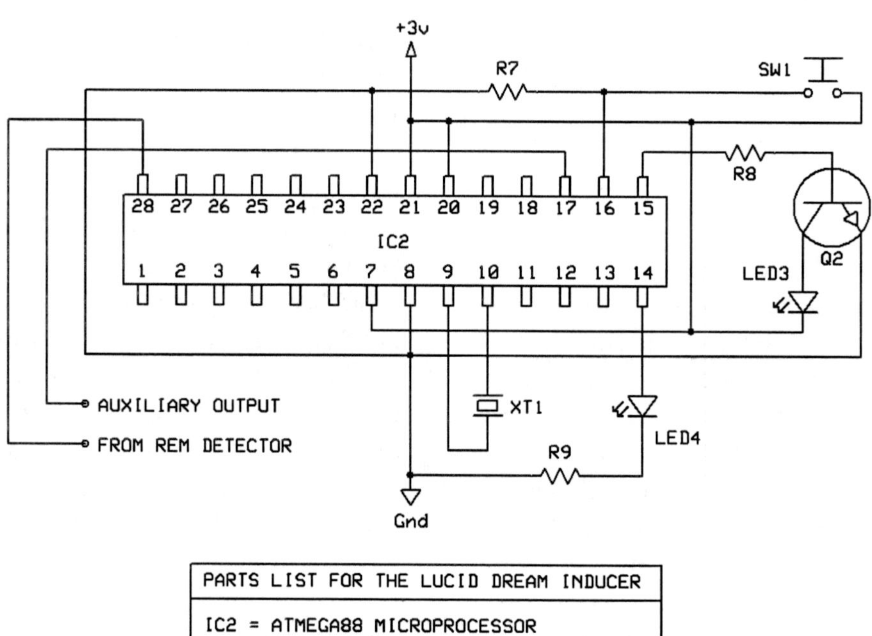

Figure 13-1 *Lucid-dream-mask controller schematic.*

Each brand of microcontroller has its own unique instruction set of assembly language, and as you progress into the hobby, you will usually end up programming in assembly so that you can create the smallest, fastest, and most efficient code possible. Of course, high-level languages such as C and Basic are much easier to start with and are good for rapid prototyping because you usually can get your program working in a few hours unless it is a real monster. For this reason, I chose Basic because it is so easy to understand that you probably could make modifications even if you have never written a program in your life. Basic is also available for most of the major brands of microcontrollers, and the language is very similar among the different varieties.

Now back to the schematic. The Atmega88 runs fine on just 3 V and only needs an external crystal to start working once your program is compiled and loaded into the internal Flash memory. XT1 is a 4-MHz quartz-crystal resonator and was chosen only because it is a common value and I had many of them in my junk box. The Atmega88 actually offers an internal 8-MHz clock source, so the crystal is not even necessary, but I have a lot of source code designed to work at 4 MHz, and the internal clock is not always the most accurate clock source.

Pin 15 is configured as an output and will drive the transistor that switches on the visible LED (LED3) that will try to signal the dreamer to become lucid. You also can try using the resistor directly from the microcontroller's output pin through a 150-W resistor because it may not need to be all that bright since your eyes are already adjusted to the darkness. I found that the system worked best if the LED was so bright that it was almost too bright when trying out test mode. The test LED (LED4) and resistor R9 are used only while breadboarding the system, so they can be omitted from the final design. LED4 will flash every time a change in input data has been detected. Also, test mode does the same thing. Pin 16 is set to be an input and tied low so that the pushbutton switch (SW1) can be detected as the pin goes high. This button controls reset as well as test mode by holding it down for 5 seconds. Pin 17 has been set up for use as an auxiliary trigger in case you want to activate some other external device during REM detection. The audio dream director presented earlier in this section would work well here. And of course, pin 18 is the input pin that will read the PWM or voltage changes from the REM-detection system. When using the infrared phototransistor, the input is fed to the onboard analog-to-digital converter, and when using the accelerometer, the pin is just a digital input pin used to measure the pulse-width time. This is why two versions of the code are presented.

Programming a microcontroller is as easy as pressing the "program" button—well, as long as your program works, because we all know the rule "garbage in, garbage out" when it comes to programming. I work with both PicMicro and Atmel microcontrollers and consider both to be equal when taking into account such things as price, support, speed, and ease of use. Of course, PIC versus AVR is like Ford versus Chevy, so I don't plan to go there! Both platforms have good C and Basic compilers available, and the assembly-language instruction set is very easy to understand. In this project I chose the Atmega88 from Atmel because it was inexpensive, fast, had more than enough IO pins, and my programmer, the STK500 (right side of Figure 13-2), allows for programming directly in circuit or on the breadboard. My PIC programmer requires me to place the chip in its onboard socket, so this is less convenient when trying out a lot of small code changes.

No matter which microcontroller brand you wave the flag for, you will need a programmer that can support the device you plan to use. Most commercial programmers will support many or all of the devices in a certain microcontroller family, and this is true with both the STK500 and the PicStart Plus programmers that I often use. There are also many low-end or home-brew programmers available, but be aware that support

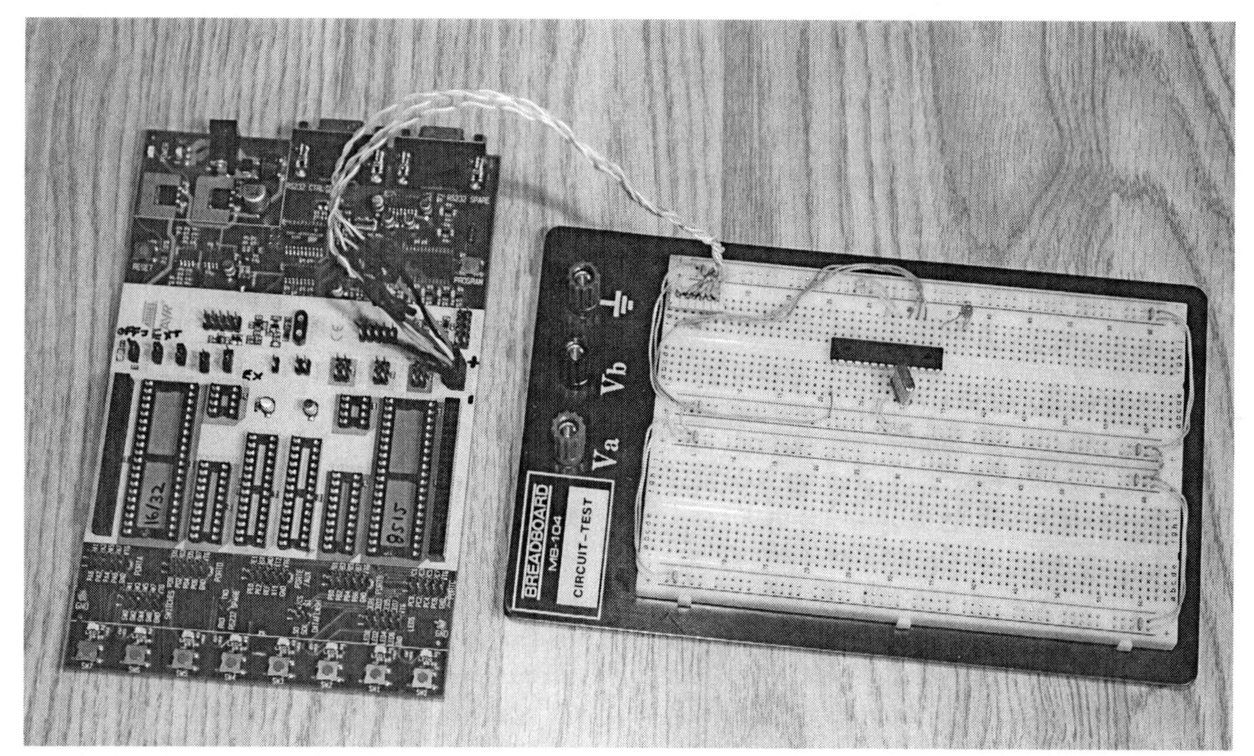

Figure 13-2 *Programming the AVR in circuit.*

may not be as good, and you may find that they may support only a limited number of microcontrollers. The STK500 shown in Figure 13-2 is probably my favorite programming board because it lets me program almost all the 8-bit AVR family, and I can do it right on the breadboard or directly in circuit. Being able to quickly alter lines of code and see the results right away, in my opinion, is a far more useful debugging process than using a software-based debugger.

Figure 13-3 shows the Atmega88 in my breadboard ready to be programmed. The four wires coming in from the left go back to the STK500 programming board, which is connected to the serial port on my computer. To program the chip, I just press the "compile and program" button on the Basic compiler, and away it goes into the Flash memory of the microcontroller. Once in the microcontroller's memory, a program is there forever, although you can make as many changes as you like. To program a PicMicro device, I have to remove the chip from my breadboard and place it in the programmer's onboard socket, a process that is fine as long as you don't plan on making a lot of changes to your code. Of course, there are in-circuit programmers available for PicMicro devices as well, but I have not used them myself.

Although I work mainly with assembly these days, I must admit that you can get a lot done in a very short time using Basic, and Bascom AVR is a really nicely done compiler with a rich set of commands and support for most 8-bit AVR devices. Figure 13-4 shows the Bascom AVR set to compile the lucid-dream-mask controller program in the Atmega88 device. Of course, when optimal speed and code size matter, assembly will kick Basic and C right out the door. For rapid prototyping and experimentation, it's hard to beat the ease and speed at which a program can be created using a high-level language. Since our dream-mask program is only reading a simple input and then running a counter, speed and code size are of no concern, so Basic is a good choice.

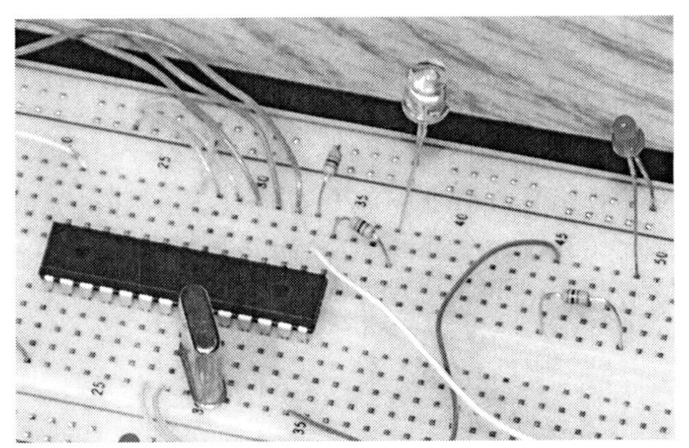

Figure 13-3 *Connecting the in-circuit programming lines.*

Now let's look through both versions of the program code so that you can see what happens inside the microcontroller. We will start with the infrared phototransistor–based version first because it was presented first in this section. Also, since much of the code is the same in both versions, this one will be explained in more detail.

Have a read through the complete Basic source code of Listing 13-1 provided in the appendix so that you can get an idea of what the code is doing. If you are an experienced programmer, then this trivial code is probably something you could write in 15 minutes from scratch, but if you have never written a program in your life, not to

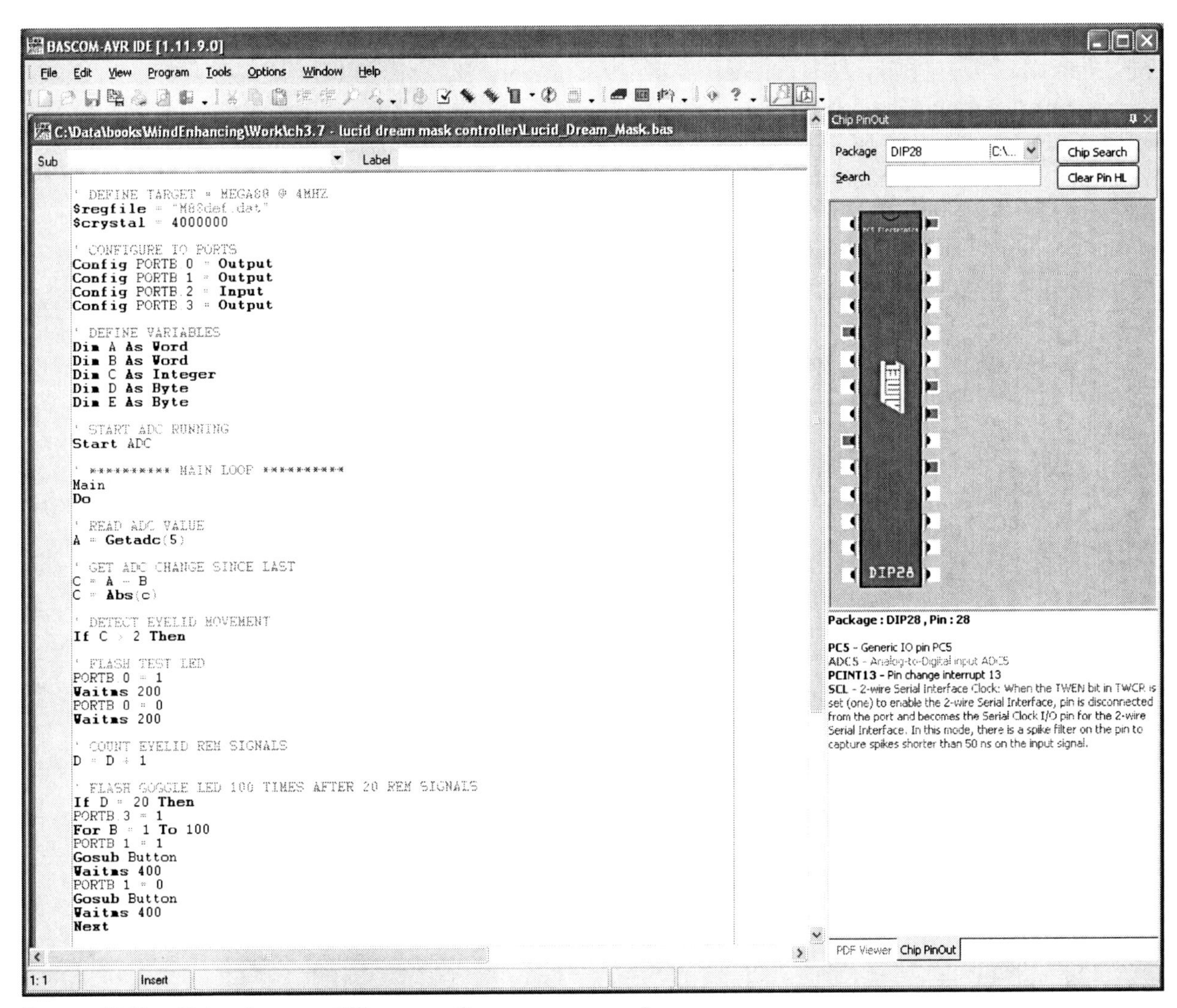

Figure 13-4 *Using Bascom AVR to compile my program code.*

worry because Basic is called that for a reason and is easy to understand. The lines in the program listing that start with an apostrophe are comments, and I will explain the code in blocks after each comment.

```
' DEFINE TARGET = MEGA88 @ 4MHZ
$regfile = "M88def.dat"
$crystal = 4000000
```

The code following this comment is required to tell Bascom that we are going to target the Atmega88 device and that our clock will be an external crystal resonator running at 4 MHz. Telling the compiler your clock speed becomes important when using commands that deal with timing-sensitive routines such as serial transmission or analog-to-digital readings. Defining the device also helps the compiler to generate user errors that have to do with IO pins. In this way, you can't accidentally try to toggle an IO pin that does not exist on the actual device.

```
' DEFINE VARIABLES
Dim A As Word
Dim B As Word
Dim C As Integer
Dim D As Byte
Dim E As Byte
```

Basic uses *variables*, which are letters or words used to hold values. I like to use single letters such as *A*, *B*, and *C* for simple programs like this one, but when you are working on a large, complex program, use of more descriptive variable names is recommended. TIMER2 and REDLED1, for example, are descriptive variable names that make a lot more sense in a huge block of code. Variables also are defined as the number of bits they are to contain, so in our code, "A" and "B" are 16-bit "words," which can contain a value of between 0 and 65535. Variable "C" is an integer that can range in value from –32768 to +32767. Variables "D" and "E" will only contain values from 0 to 255, so they are bytes. Although you could just define all variables using larger data types, this is a waste of memory space and will slow down your code.

```
' START ADC RUNNING
Start Adc
```

The "Start Adc" command tells the compiler to include the code necessary to set up and initiate the onboard analog-to-digital converter on the Atmega88. This will allow us to read in an analog voltage and convert it into a value in order to test the state of the phototransistor on the dream mask.

```
' ********** MAIN LOOP **********
Main:
Do
```

Everything from here on is going to happen continually until the word "Loop" is reached, which causes program execution to start again where it first encountered the word "Do." This is called an *endless loop* because it never stops unless forced to by another command or an error.

```
' READ ADC VALUE
A = Getadc(5)
```

This command reads the analog-to-digital converter (ADC) on pin 5 of the Atmega88 into variable "A," which is where the output from the phototransistor is connected. Since the ADC returns a 10-bit reading, values can range from 0 to 1024, which is why variable "A" needed to be a "Word," not a byte.

```
' GET ADC CHANGE SINCE LAST
C = A - B
C = Abs(c)
```

These next two lines compare the current ADC reading "A" with the last known reading "B" by subtracting them. The "Abs(C)" command changes the value in "C" to an absolute nonnegative value, so we get the difference only as a whole number, not a negative number, if "B" is greater than "A."

```
' DETECT EYELID MOVEMENT
If C > 2 Then
```

This line checks to see if the value in variable "C" is greater than 2. If it is, then all code following the "Then" statement will be executed. If it is not, then the program will skip ahead until if find a line the reads "End If." In this way, the counter processing code to follow will not run unless the reading on the ADC has changed by a value of at least three out of a possible 1024 values. Thus, if you want to reduce the sensitivity of your dream mask, just increase this number.

```
' FLASH TEST LED
Portb.0 = 1
Waitms 200
Portb.0 = 0
Waitms 200
```

If there is an LED connected to pin 14 (Portb.0) of the Atmega88, it will blink for 200 ms once a change of greater than 2 has been detected from the ADC. The "Waitms" command simply delays code execution for any number of milliseconds, so the LED goes on for 200 ms and then goes off and waits for another 200 ms. Writing a 1 to a port sends VCC to the port, whereas writing a 0 to the port clears it to ground.

```
' COUNT EYELID REM SIGNALS
D = D + 1
```

"D" is a variable that will hold a counter that counts the number of times the ADC has detected a change. Since we don't want the dream mask signaling us every time we twitch an eye, we use a counter to keep track of eyelid detections so that we can wait until they are happening nonstop.

```
' FLASH GOGGLE LED 100 TIMES AFTER
      20 REM SIGNALS
If D = 20 Then
Portb.3 = 1
For B = 1 To 100
Portb.1 = 1
Gosub Button
Waitms 400
Portb.1 = 0
Gosub Button
Waitms 400
Next
```

This block of code will execute only if the counter variable "D" reaches 20 so that we can basically filter out any spurious eye movements.

Once "D" reaches 20 counts, pin 15 (Portb.3) is turned on and off 100 times at a rate of 800 ms. Pin 15 is connected to the base of the LED diver transistor, so this will switch it on or off. During each LED flash, the "Gosub Button" command jumps to the button-handling routine just to make sure that we are not trying to reset the device in case this was a false detection owing to eye twitching or an accidental wake-up from REM sleep. If you find that there are too many false detections, then just increase the number 20 to something higher. Also notice that "Portb.3=1" sets pin 17 to VCC, which is an auxiliary output in case you want to connect some other device to the unit.

```
' RESET AND START OVER
Portb.3 = 0
D = 0
End If
End If
```

After flashing the indicator LED in the dream mask 100 times, the program simply resets the variables, turns off the auxiliary pin, and then starts over again from the beginning. If you have another REM episode, the program will again detect it and flash the indicator LED in your dream mask. The two "End If" lines close off the previous two "Then" statements.

```
' RESET LAST ADC VALUE
B = A
```

This line simply makes the variable "B" take on the last value of the ADC so that when a new value is read for "A," it then can be compared with the old value to see if it is different.

```
' CHECK FOR BUTTON PRESS
Gosub Button
Loop
End
```

Again, the program calls the button-handling routine in case the user is pressing the reset/test button. "Loop" then sends the program back to the "Do" statement so that the entire loop can continuously run.

```
' ******** BUTTON PRESS ********
Button:
If Pinb.2 = 0 Then Return
```

This is the start of the button-press routine, and the first line checks to see if the button is actually pressed by reading pin 16 (Pinb.2); otherwise, it just returns back to the calling line.

```
' RESET SYSTEM
D = 0
B = 100
E = 0
```

If there was a button press, some of the variables are cleared, and "B" is set to 100, which will cause the LED flashing counter to end if it was running, once the button-handling routine is over.

```
' ENTER TEST MODE
While Pinb.2 = 1
Waitms 20
E = E + 1
```

```
If E = 200 Then
E = 0
Portb.0 = 1
Goto Setup
End If
Wend
Return
```

Because things happen in megahertz in a microcontroller, we need to use counters and delays to compute seconds. Variable "E" increments as the user holds down the pushbutton, and then the code delays for 20 ms. If the user holds the button down long enough for "E" to count up to 200, then the program assumes test mode because about 5 seconds have elapsed. The routine then jumps to the "Setup" label in the code to run setup mode. If the button was not held for at least 5 seconds, then the routine just ends and returns back to where it was called.

```
' ********* SETUP MODE *********
Setup:
Do
```

This is the beginning of setup/test mode, and it is another endless loop.

```
' GET ADC VALUE
A = Getadc(5)
C = A - B
C = Abs(c)
If C > 2 Then
```

Again, the program reads and compares the current ADC value "A" against the last known value "B" to see if there was a change of greater than 2 of 1024 values.

```
' SETUP MODE LED FLASH
Portb.1 = 1
Waitms 50
Portb.1 = 0
Waitms 100
End If
B = A
```

If an ADC comparison exceeds a value of 2, then this code block flashes the visible LED so that the user can see the results directly as he or she keeps his or her eyes closed and moves them back and forth. In this way, you can tell if the system is functioning properly or not in real time.

```
' EXIT SETUP MODE
While Pinb.2 = 1
Waitms 10
E = E + 1
If E = 200 Then
Portb.0 = 0
E = 0
Goto Main
End If
Wend
Loop
```

This block of code is much like the block of code that was used to enter setup mode. This time, if the user holds down the pushbutton for 5 seconds, the program exits setup mode and returns to the main loop to go back to normal operation.

Now you can see why Basic is great for rapid prototyping; the language is almost pure English and is so easy to work with. The complete accelerometer version of this code, shown in

Listing 13-2 and provided in the appendix, is very similar, and since there are only a few changes, I will just post the differences and explain them.

Since there are no analog voltages to deal with, the ADC is not used this time, so the obvious difference will be the exclusion of "Start Adc" in the code.

```
' ********** MAIN LOOP **********
Main:
Do
' READ ADXL202 VALUE
Pulsein A , Pinc , 5 , 0
```

Instead of reading the ADC value, we read pin 5 as a digital input, but measure how long it takes for it to go from high (VCC) to low (GND) and then back to high again. The "Pulsein" command does this for us automatically. This length is stored in the variable "A" just like the ADC reading was in the preceding version of the code. Because of this, most of the code can be reused because we really don't care what the value is but only that it has changed since the last reading.

```
' ********* SETUP MODE *********
Setup:
Do
' GET ADXL VALUE
Pulsein A , Pinc , 5 , 0
C = A - B
C = Abs(c)
If C > 2 Then
```

Of course, in setup mode, we also have to remove the ADC reading and replace it with the "Pulsesin" command, but other than that, the two versions of the code are exactly the same.

There is a lot of room for further improvements to this code and plenty of program space as well as IO pins left on that microcontroller, so get creative and add your own neat features and options. A much better REM-detection routine would include a timeout counter so that rather than just counting eyelid movements, the counter resets after a few seconds if eyelid movements stop. This would greatly reduce false detections from erroneous movements or eye twitches. You also could add a sound-alert system right into the microprocessor using one or more of the output pins to make beeps or complex noises or even play sound digitized from an external EEPROM memory. The possibilities for modifications and improvements are endless, but I only had so much space to add code to this book, so the "bells and whistles" are up to you.

There is also an option to use a microcontroller for this project that will allow you to collect the data into your PC for processing and analysis. There are many external ADC devices such as the one shown in Figure 13-5 that will read in an analog value and then transmit it to your PC through the serial or USB port.

The unit shown in Figure 13-5 is the Pico ADC from Picotech, and it is a very accurate and inexpensive ADC that can be used with the included data logger or in your own programs. Drivers for this ADC are even included for Visual Basic, so you can write a program almost exactly the same as the Atmega88 program to test your dream mask. Of course, with a computer, you have to run wires from the mask, but you open up a whole new bag of possibilities by using your PC to trigger your dreaming mind. You could add sound files or connect LEDs, buzzers, or even tactile feedback hardware to your body for some very interesting experiments. Of course, you will now be plugged into the ac system, so again, the risk of a lightning hit or power surge is always

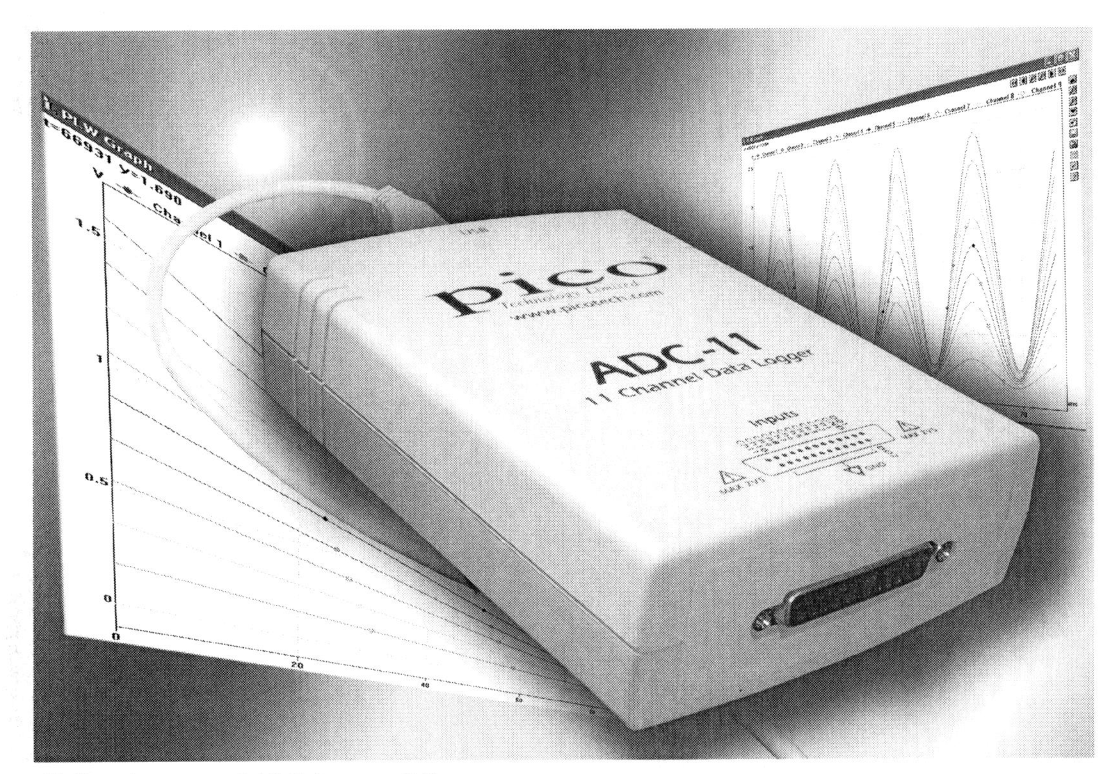

Figure 13-5 *An external ADC for your PC.*

there. There are many rf modules available that can transmit and receive data for short distances, so you might want to consider that option if you decide to use your PC in your "sleep laboratory" at night.

Now you have a powerful tool to aid you in your lucid-dreaming journeys, so have fun experimenting with it, and remember that the dream mask alone may not be enough to induce a lucid dream. You need to practice proper dream-recall techniques and train yourself to do a reality test anytime you see a bright light or flashing light source in the real world. These could be car signal lights, a video game screen, room lights going on and off, and the like. Once you begin to automatically ask yourself, "Am I dreaming?" each time you see a bright or flashing light, you eventually will begin to do the same in REM sleep as the mask flashes its visible LED for you. The instant you realize that you are indeed dreaming, you can become an oneironaut and take control of your dream world!

Remember the waking reality tester presented earlier in this chapter? Well, why not remove the vibrating motor and modify it to flash a bright LED instead? The unit then could be placed near your workstation and help you get into the routine of seeing the flashing light signs, as well as remind you to question your reality at the same time.

There are many more techniques you can use to help your dreaming mind into a lucid state, so have fun, be patient, and dig around on the Internet for more information if you enjoy this subject; there is a lot of great information out there. Maybe I will see you in a dream someday. Oh wait, that's an entirely different "fringe science" for another time! In Section Three we will go on an inward journey using meditation, clairvoyance, hypnosis, and color-therapy devices.

Section Three

An Inward Journey

This section explores the amazing biological computer that we all carry around with us, our brain! Unlike the art of hacking a computer, you can't simply jack a few wires onto the CPU or alter the program code that makes us who we are, well at least not in any safe manner, so we will explore a few interesting methods that will allow us to "talk" to parts of the brain associated with varying levels of consciousness. Plugging a jack into the back of your spine in order to enter a large virtual world like Neo did in the Matrix is something that will certainly be a reality some day, but for now we must work with the simple, nonevasive tools we have available, and become pioneers in this interesting field. Who knows, maybe one day you will be the one to develop the technology that will allow us to project our consciousness into a virtual world! Of course, we have to start with the basics, so get out your soldering iron, and let's have some fun!

Project Fourteen

The Ganzfeld Effect

In the 1930s, Wolfgang Metzger, a German psychologist, found that if the eyes are deprived of any focal point or depth of field, the brain basically "disconnects" from the visual system, causing a drastic change in our mental state. The word *Ganzfeld* is German for "complete field" and has been the subject of much investigation into meditation, hypnosis, and even parapsychology, which includes telepathy. Some reports suggest that the use of a Ganzfeld effect device can bring about the same mental state as achieved by those who have practiced meditation for years.

The Ganzfeld effect is also the same effect reported by those who have experienced snow blindness, which often leads to strange hallucinations and a general altered state of consciousness. Some of the most profound research into the Ganzfeld effect was in the area of telepathy, where it has been shown that using a Ganzfeld device can greatly enhance the probability of a successful telepathic transmission. In these experiments, a "receiver" is often in a secluded room using a Ganzfeld device (such as the one that will be presented here) and is allowed to relax for some time so that sensory deprivation occurs. The "sender" then mentally focuses on some random image and attempts to send it telepathically to the receiver. Typically, the average "hit rate" of such an experiment would be around 25 percent, but by using the Ganzfeld device, this rate was shown to be over 35 percent much of the time, even by some researches who do not consider telepathy to be possible.

Since the Ganzfeld device is such a simple contraption to build, why not conduct your own telepathy experiments, or just use the device to send yourself into an altered state of consciousness somewhere between waking reality and the dream world? All you will need for parts are a few Ping-Pong balls and a handful of red light-emitting diodes (LEDs) to build this simple project.

The main component you will require is not a sophisticated IC or a hard-to-find semiconductor, but just a pair of ordinary white Ping-Pong balls. Most of the research into the Ganzfeld effect was done using Ping-Pong ball halves placed over the subject's eyes and a red-light source to create even lighting. The goal is to completely remove all points of reference from the subject's field of view so that the eyes cannot focus on anything, not even a change in brightness or color, which is why a light is usually used. In most of the experiments, red light is used, but since the Ping-Pong balls are flat white on the inside, you can illuminate them with any color light, or just find

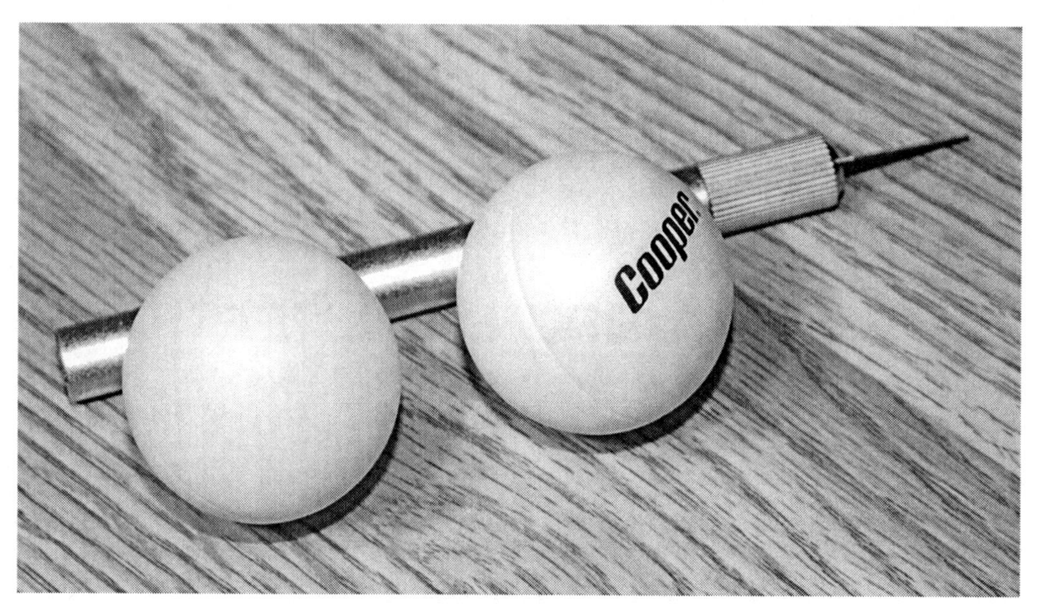

Figure 14-1 *Ping-Pong balls with a logo on only one side.*

a comfortable chair outdoors and use sunlight for illumination.

As for the Ping-Pong balls, there is nothing special about them, but you will need to find a set with a logo on only one side, as shown in Figure 14-1, so that you can cut them in half and use the featureless halves.

There will be an equator along the circumference of the Ping-Pong ball dividing the logo side from the blank side. Cut a slit on the logo side as shown in Figure 14-2 using a steak knife or razor knife just large enough to insert the tip of a small pair of scissors. The smaller the scissors, the better because the plastic is very thin

Figure 14-2 *Cut a slit on the logo side of the equator.*

and brittle, so you will want to make small cuts along the edge of the seam to cut the ball evenly.

Try to make small cuts using only the scissor's tip as you work your way around the equator, as shown in Figure 14-3. This will help to avoid the sharp sawblade edges that happen if the cut becomes misaligned. You also could cut around the ball with a Dremel tool if you have the tiny grinder disk attachment.

Try placing the ball halves over your eyes to see how they fit. I found there to be a tiny gap near the sides of my head where the edges did not conform to the shape of my face, and this would greatly reduce the effectiveness of the sensory-deprivation experiments. The goal is to see nothing that you can focus on, so you may have to cut a small slice out of each side of the ball halves as shown in Figure 14-4 to help keep out any stray light or viewable edges.

I cut a bit out of the nonusable ball half just to see what shape would work best for conforming to my face and then used a marker to trace it onto the good halves so that they could be cut. As shown in Figure 14-4, this required taking an arc about half an inch into each edge so that the ball halves made a better seal with my face. You may find that this is not necessary, but you shouldn't have to press down on the balls to create a good fit because this would hamper your ability to relax during the experimentation.

The schematic shown in Figure 14-5 is a guide that you can use to figure out how to run multiple LEDs from a single 9-V battery without the need of a regulator. All you have to do is divide your source voltage (9 V in this case) by the rated voltage of your LEDs (2.3 V in my case). Thus, since 9 divided by 2.3 is 3.9, I can safely run four LEDs in series, which will give each one 2.25 V. It's always better to be slightly lower than the rated maximum than slightly higher if you want your LEDs to enjoy a long, healthy life. Notice that for each series arm, I have four LEDs also wired in parallel. You can have as many LEDs as you want in parallel as long as your power source can handle the current draw. To sum up, then, series wiring divides the voltage by the number of LEDs in series, and parallel wiring increases current draw by the number of LEDs in parallel. In my configuration, I used four LEDs in series

Figure 14-3 *Cutting the ball along the seam.*

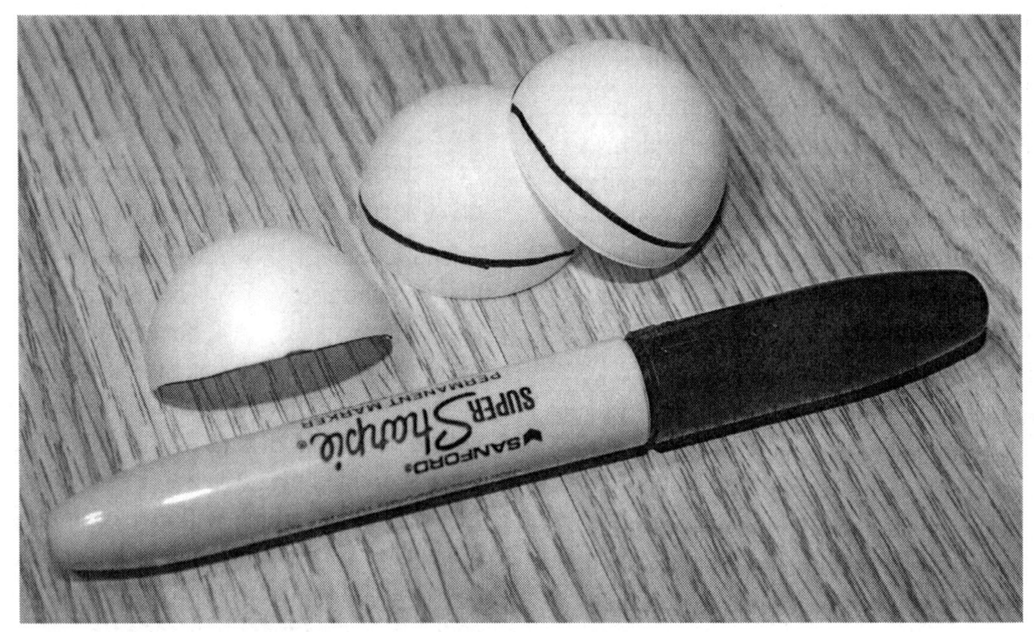

Figure 14-4 *Shaping the ball halves to conform to your face.*

with four parallel legs so that there were eight LEDs for each eye—16 in total.

The series and parallel LED configurations are shown on the breadboard in Figure 14-6 as I tested the brightness using a 9-V battery. Remember, when adding LEDs in series, it's best to end up under the maximum voltage than slightly over because many LEDs cannot tolerate overvoltage. Also, you may need fewer than eight LEDs for each eye if you can find a more diffused lens style. The ones I used were the ultrabright type and were highly focused to a narrow field of view, which is why they required further diffusing to avoid hot spots.

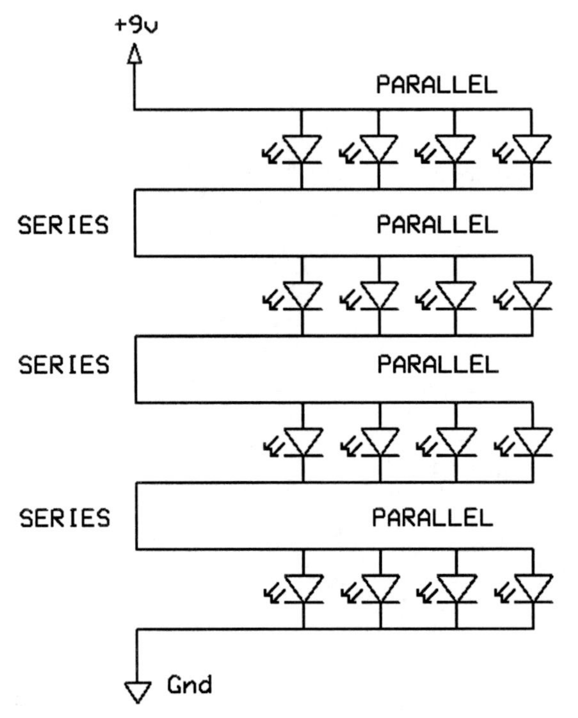

Figure 14-5 *Parallel and series LED wiring.*

If your LEDs are highly focused and very bright, then they may cast hot spots and shadows on your eyepieces when in use. The best way to find this out is to just light them up and place them about 4 to 6 in above your eyes while you wear the Ganzfeld eyepieces. You want a bright light source, but not one that will create spots that you can focus on. To filter out the hot spots, just cut up another pair of Ping-Pong balls and place them a few inches above the LED cluster to diffuse the light. Figure 14-7 shows the diffuser in action, and although the figure shows obvious light and dark areas, these were not visible while wearing the eyepieces and are mainly due to the way the camera took the photograph. White

Figure 14-6 *Testing the parallel and series LED hookups.*

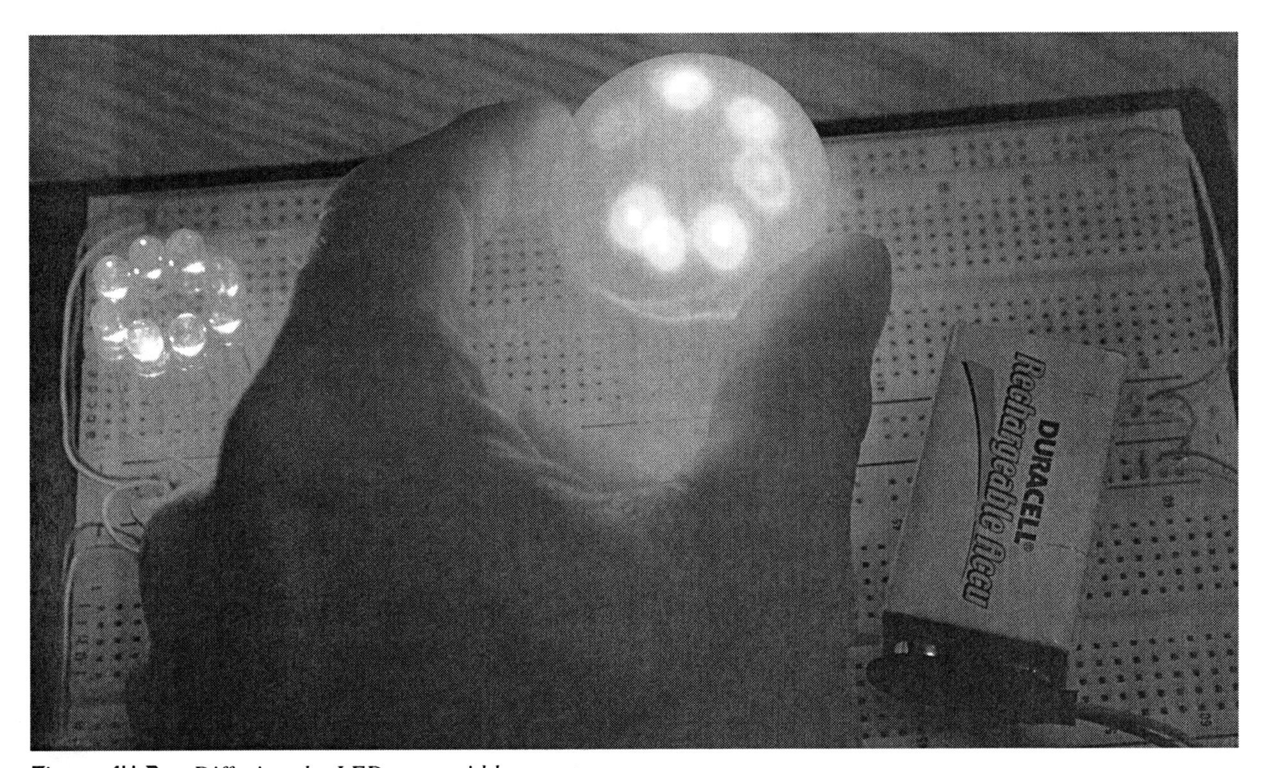

Figure 14-7 *Diffusing the LEDs to avoid hot spots.*

Figure 14-8 *Soldering the LED clusters to a perf board.*

Styrofoam and even a sheet of white paper also will work well as an LED diffuser.

Once you have figured out how many LEDs you will need, they can be mounted to a bit of perf board for later installation into a mounting tube of some type. Figure 14-8 shows my dual LED clusters after following the same wiring scheme used on the breadboard when testing them. In case you do not already know, LED polarity can be identified either by the longer lead (positive) or by the flat part on the housing (negative). This can come in handy if you have cut the leads short and find that one or more of your LEDs fail to light up when power is applied.

Since I wanted to be able to place the two LED lighting units over my head in various positions as I wore the Ganzfeld eyepieces, some type of housing was needed that could attach to a tripod for easy adjustment. I found a bit of PVC tubing, as shown in Figure 14-9, to fit the Ping-Pong ball halves perfectly, and two lengths were cut at the correct focal length for optimal light diffusion. At 5 in long, the LED light hits the inside of the Ping-Pong ball diffuser at an optimal angle so that no hot spots were present when pointing the light over the Ganzfeld eyepieces.

As shown in Figure 14-10, both the ball halves and the LED clusters were secured to the PVC

Figure 14-9 *Making housing for the components.*

Figure 14-10 *A hot-glue gun is an essential tool.*

tube using a hot-glue gun. A hot-glue gun is one of those "must have" tools for every electronics hobbyist and will create not only a secure bond but also one that can be broken if needed. Remember to drill any holes you need for wiring or other mounting hardware into the PVC tubing before you glue everything together. Besides PVC tubing, practically anything can be used to position the diffuser over the LED clusters. Cardboard tubes, plastic boxes, or even an old shampoo bottle washed out and dried thoroughly would work just fine for this job.

To make a simple mounting system that would attach to a tripod for easy adjustment, I cut a 2-ft length of PVC tubing as shown in Figure 14-11 so the two LED units could be mounted on either side. The diameter of the PVC tube was perfect to allow the distance between the two LED units to be the optimal distance for my eyes. Using a tripod to position the light source made it easy to find a comfortable place to use the Ganzfeld system without worrying about setting up the hardware. The tripod allows height adjustments from 2 ft to over 6 ft off the ground.

The completed light source is shown mounted to a camera tripod in Figure 14-12, ready to use with the Ganzfeld eyepieces. For extended periods of use, you may find it more convenient to use a 9-V direct-current (dc) adapter pack rather than the battery because multiple LEDs can become power-hungry, especially the high-brightness types. I found that a fully charged 9-V battery would run my lighting system for about an hour before it began to fade, so that was okay for short experiments where I did not plan to get into any really deep altered states. If you intend to really "go deep," then you most likely will have to switch to a dc adapter.

Considering that there is not much in the way of hardware to this project, it really does have some profound effects on a person's state of consciousness. Figure 14-13 is unfortunately lacking in color, so you cannot see that the red glow from the LED lighting system was very rich as I attempted to reach higher levels of consciousness. If you are interested in meditation or hypnosis, then the Ganzfeld device can be a powerful aid in reaching a trancelike state.

Figure 14-11 *Adding a support tube for tripod mounting.*

Figure 14-12 *The completed tripod-mounted light source.*

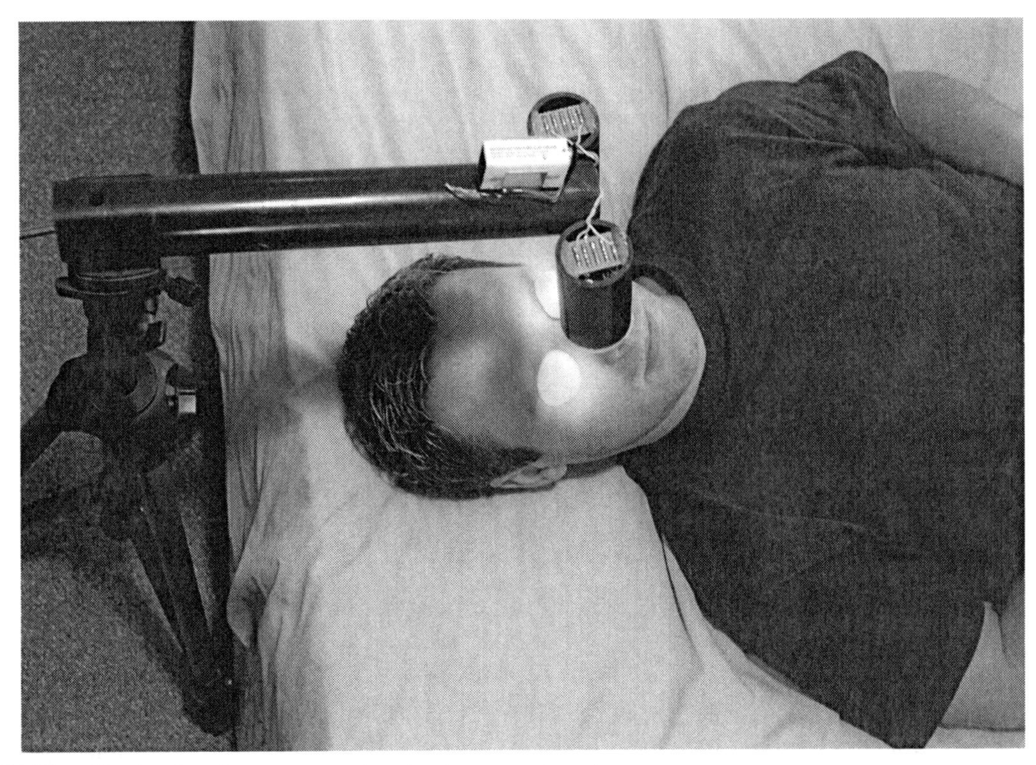

Figure 14-13 *Getting lost in a sensory-deprivation experiment.*

Maybe you want to experiment with telekinesis or PSI, attempting to recreate some of the well-known experiments done using hardware almost identical to what you have just built. How about trying some remote-viewing experiments? A *remote viewer* is a person under a trancelike state who can see a remote location as if viewing it through some type of invisible camera link. This may sound far out to you, but there must be something to it because the U.S. and Russian military have invested millions into remote-viewing research, often performing experiments with Ganzfeld devices. Further reading on the military's research can be found by searching for the StarGate Project on the Internet.

Well, I do hope that you enjoy your altered states, and if you discover some untapped power by accident, be sure to let me know. Oh, and don't use your new psychic tools and powers for anything too devious!

Project Fifteen

Alpha Meditation Goggles

Depending on our mental state, our brain is constantly oscillating at various frequency ranges. These oscillations, or *brain waves*, occur when numerous groups of neurons become coherent in their electrical activity. Alpha wave is the name given to the range of brain-wave oscillations that fall between 7 and 12 Hz (cycles per second). *Alpha waves* are also known as Berger's waves in memory of *Hans Berger*, who made the first known electroencephalographic (EEG) recording of a human subject.

Alpha waves are present when we are in a relaxed state, often with our eyes closed, and bring about a heightened feeling of well-being. Interestingly, the earth's resonant frequency (also known as the *Schumann resonance*) happens to fall within the alpha-wave frequency range at 7.83 Hz. Some people have speculated that our minds would be "in tune" with the earth at this frequency.

The device presented here will let you cover the entire alpha-wave frequency range so that you can find your "magic" frequency when trying to relax or meditate. The purpose of the alpha meditation goggles is to coax your brain into synchronicity with the alpha-wave frequency of the flashing LEDs in order to reach a very relaxed state in a short time.

Please note that flashing lights (and video games) have been known to trigger epileptic seizures in a very small percentage of the population who have a rare form of epilepsy that seems to respond to external visual stimulus. Although extremely rare and only a risk to those who are known to have epilepsy or have not yet been diagnosed, it is worth noting this precaution before you build and use any device with bright flashing lights.

The schematic shown in Figure 15-1 will pulse the two or more LEDs that will be mounted to your goggles or mask. The 555 timer IC is working as an oscillator or pulse generator with its rate controllable via variable resistor VR1. You also can adjust the brightness of the LEDs by moving variable resistor VR2 to control the voltage to transistor Q1, which drives the LEDs. The working range of the oscillator can reach well below and well above the alpha-wave frequency, so you even could experiment with the other brain-wave groups as well, although alpha waves are the ones that are most useful in meditation and relaxation.

The schematic is very simple, but it is always a good idea to try things out on your solderless breadboard before warming up the soldering iron. Figure 15-2 shows the oscillator running on a breadboard and powering two LEDs. I found that

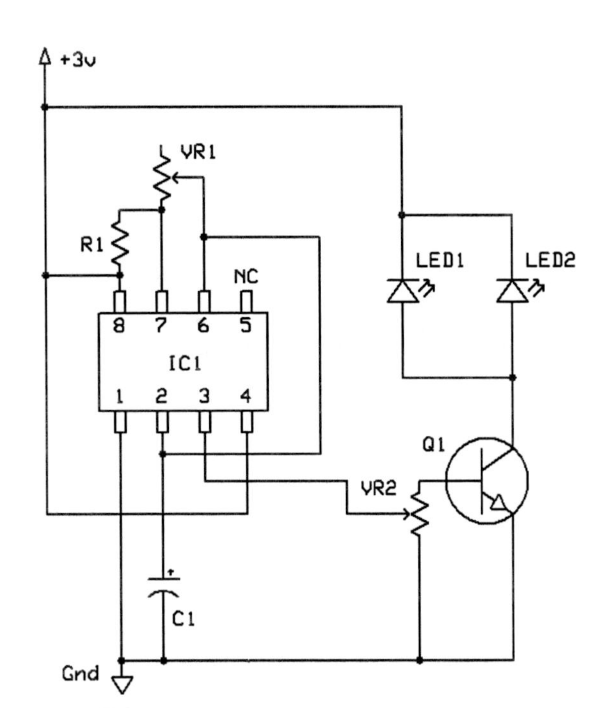

Figure 15-1 *The alpha meditation device schematic.*

PARTS LIST FOR THE ALPHA MEDITATION GOGGLES
ICI = NE555 TIMER
Q1 = 2N3904 OR SIMILAR NPN TRANSISTOR
R1 = 1K
C1 = 10UF
LED1,LED2 = HIGH BRIGHTNESS LEDS
VR1, VR2 = 10K VARIABLE RESISTOR

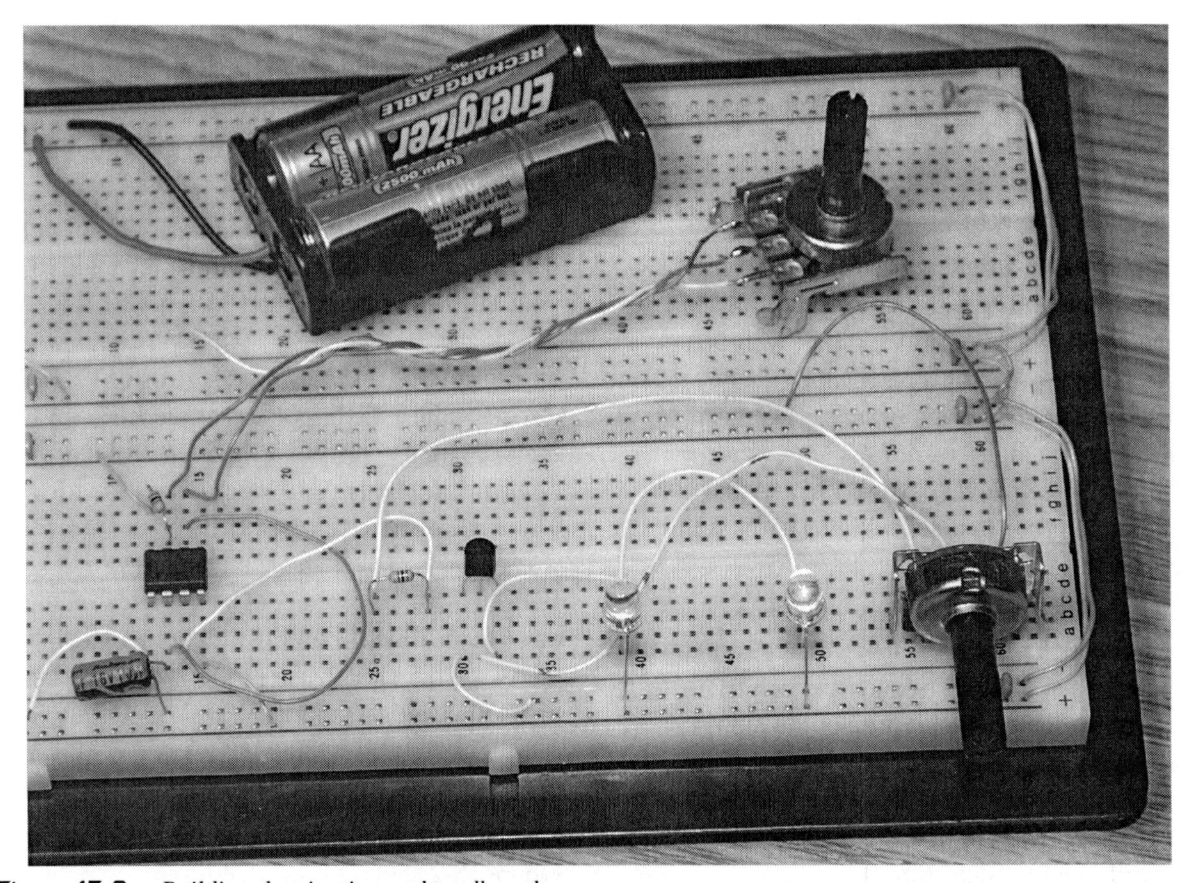

Figure 15-2 *Building the circuit on a breadboard.*

six LEDs (three for each eye) were best with the ones I was using because they could be set at a lower brightness level and cover more area that way. You will have to experiment with your LEDs, but do make sure that they are all the same make and model so that one side is not brighter than the other. The circuit is designed to run at 3 V to make battery power easy, but the 555 timer IC would be happy all the way up to 12 V. For operation above 3 V, you certainly would need to add resistors (150 Ω or more) in series with your LEDs in order to reduce the current.

Since the oscillator has no readout, you will need to use a frequency counter or oscilloscope to set the frequency into the alpha-wave range between 7 and 12 Hz at least once before use. You could just set your oscillator for 10 Hz, which is basically in the center range or experiment with the completed goggles and find your preferred frequency or the one that makes you feel most relaxed. For me, 10 Hz seemed to work just fine, but if you want the ability to adjust this at any time, then just make VR1 accessible as a control when you build the circuit into a cabinet. You also can feed an extra pair of wires out of pin 3 of the 555 timer and ground so that your frequency counter always can be connected when you want to know the actual working frequency. Figure 15-3 shows my circuit oscillating at 10 Hz, as reported by my basic multimeter.

There are many ways to get the light over your eyes, and using an old ski mask or a pair of swim goggles is a good method. The alpha meditation system is not intended for long-term use, but you still will want a setup that is comfortable and has proper air circulation, so the goggle straps are best replaced by loose elastic and should have small holes drilled in them to let the air flow freely. I decided on using the small swim goggles shown in Figure 15-4, and the straps were so long that they were easy to adjust loosely. You also may have to experiment with the type and number of LEDs that work best over your eyes so that you can achieve maximum light without having the light source be so bright that it is

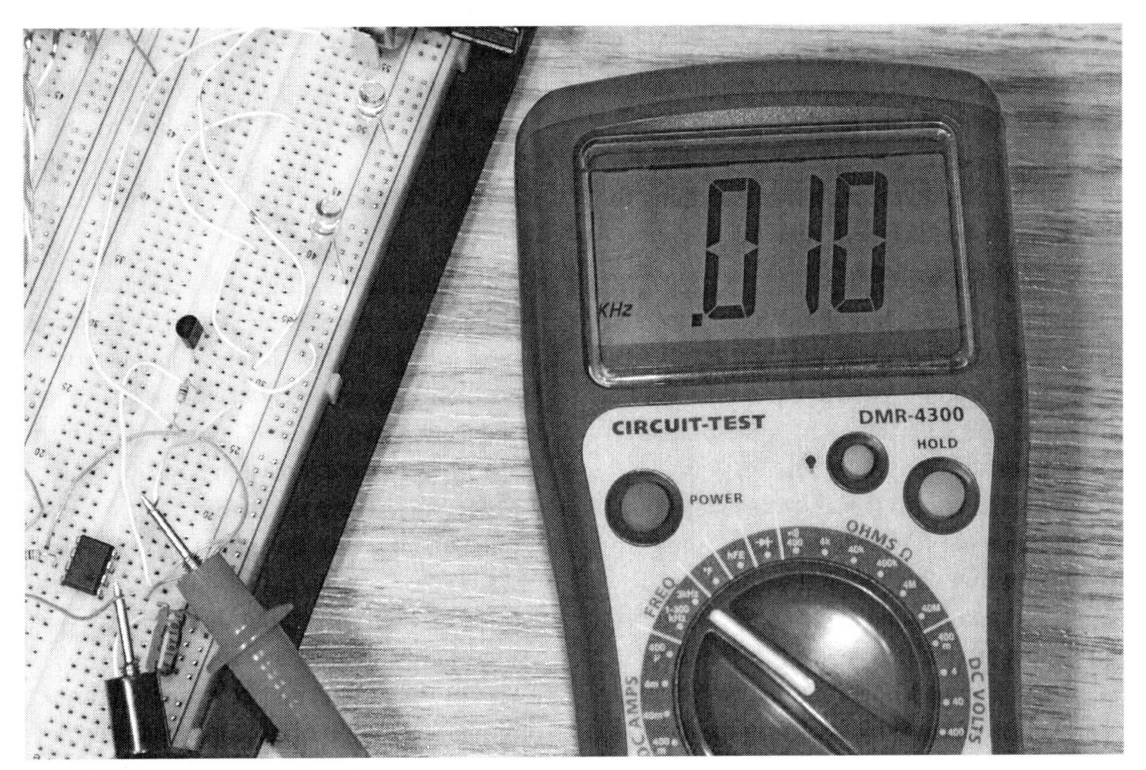

Figure 15-3 *Finding your magic frequency.*

Figure 15-4 *Choosing LEDs and a mounting system.*

uncomfortable. Your eyes should feel saturated with light, but not so much that it feels like a camera flash going off in front of your face. I found that three LEDs in each eye were good, and I still had room to turn up the brightness adjustment, if needed. A hot-glue gun is a handy tool that can be used to mount almost anything to anything else, so that is how I intended to secure the LEDs to the goggle lens.

The number of LEDs, brightness, and color all will have an effect on the way the light is delivered through your closed eyes and the lens (if there is one) on your mounting system. Red LEDs will transmit more light through your closed eyelids than any other color because your eyelids act like a red filter when closed. Any color will certainly work, but red seems least distracting and natural, and since color therapy is not part of this project, red seems like the most logical choice. Figure 15-5 shows the two small three-LED clusters made by soldering the three LEDs in parallel to a small bit of perf board. In this circuit, all LEDs are in parallel, and you can have as few as two or as many as your transistor can source current to. As you exceed your driver transistor's capacities, the LEDs will become less bright, and overheating of the transistor could occur. Of course, you could just use more transistors to drive more LEDs, but be aware that this also will shorten the life span of a battery charge.

Using the hot-glue gun, the LED clusters are affixed to the center of the goggle's lenses, as shown in Figure 15-6. Some goggles may not have their lens centers over your eyes' centers, so some testing may need to be done first. If you put on the goggles and look into a mirror, you can make a mark where the exact center for your eyes will be and use that as a guide. A nice side effect of using the hot glue is that it creates a bit of a diffuser to help spread out the light evenly when using high-brightness and highly focused LEDs. This diffusing effect can be exploited further by first melting a blob of glue over the lens before gluing the LED clusters down, further adding to the diffusing effect.

Figure 15-5 *Choosing LED color and quantity.*

I decided to add a jack to the goggles, as shown in Figure 15-7, so that they could be used for other experimentation and connected to any device capable of driving them. A ⅛-inch mono headphone-style jack simply was soldered right to the small perf board, and a matching male plug was added to the end of the wire coming from the oscillator unit.

The small circuit board, battery pack, switch, and variable resistor are mounted into a more permanent home, as shown in Figure 15-8. I decided to leave my system set for 10 Hz, so there is only one cabinet-mounted variable resistor, the one that controls LED brightness. A small trimpot variable resistor is added to the board and preset to the desired frequency before placing all the parts in the enclosure. If you plan to experiment with frequency adjustment, then just add a second cabinet-mounted variable resistor.

With your alpha meditation goggles ready for use (Figure 15-9), find a quiet and comfortable place to lie down and reach the height of Zen! Okay, it may take you a bit of practice and patience to reach that level of meditation, but synchronizing your brain waves into the alpha state will surely help you along the way. If you are planning to use the system in a dark room, then let your eyes adjust for about 15 minutes before using the goggles so that you can set the brightness for your already-adjusted eyes. The light should be bright enough to feel "mesmerizing" but not so bright that it feels uncomfortable in any way.

Looking like the cover of some cheesy sci-fi movie, Figure 15-10 demonstrates how our mind-hacking projects can be entertaining not only to the user but also to the onlookers! Flooded in a

Figure 15-6 *Mounting the LEDs to the goggles.*

Figure 15-7 *Wiring the LEDs for external access.*

Figure 15-8 *Putting together the oscillator unit.*

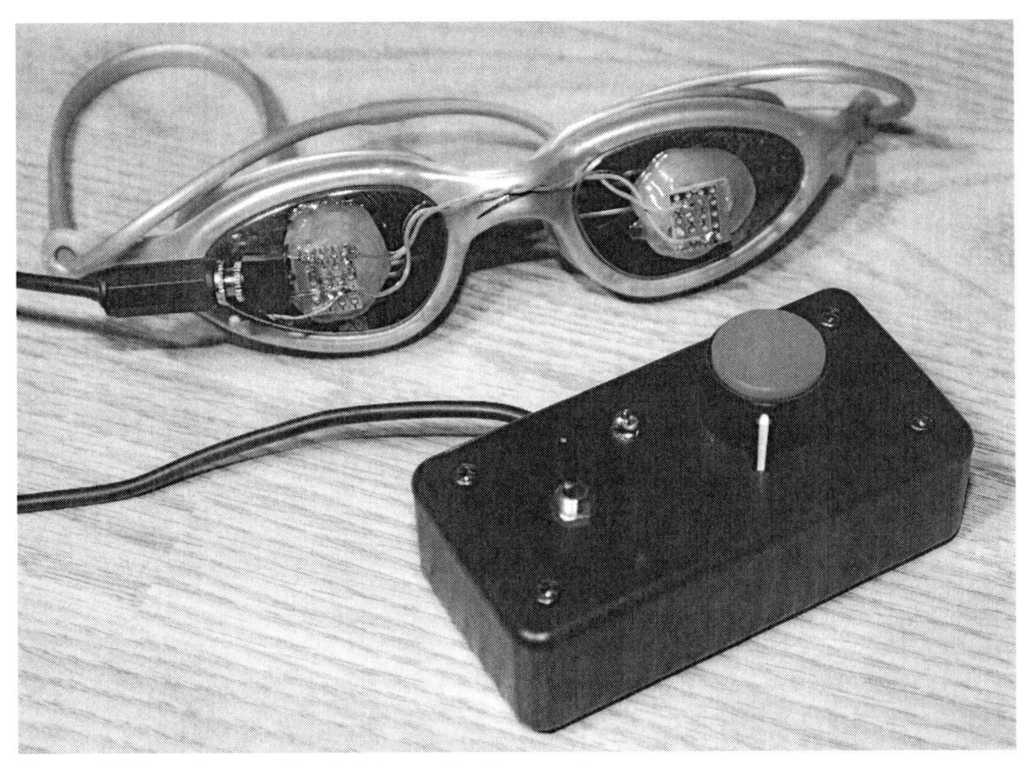

Figure 15-9 *The completed alpha meditation goggles.*

flickering pool of red light, I relax and drift into a state that feels much like a land between waking reality and sleep, using the device to power nap during the day. The alpha meditation goggles certainly do help when it comes to feeling relaxed, and there is a definite feeling of relaxation after a bit of use.

Some other experiments you might want to try with this device include the addition of sound to the output, using two oscillators to drive each eye independently, or the addition of some type of alternating color to the light source. Sound pulses can be added simply by placing a set of headphones onto the output of the LED driver transistor through a 1K resistor. Each time the LED pulses, a beat will be heard in the headphones. In the next project, we will transcend further to test our clairvoyance abilities.

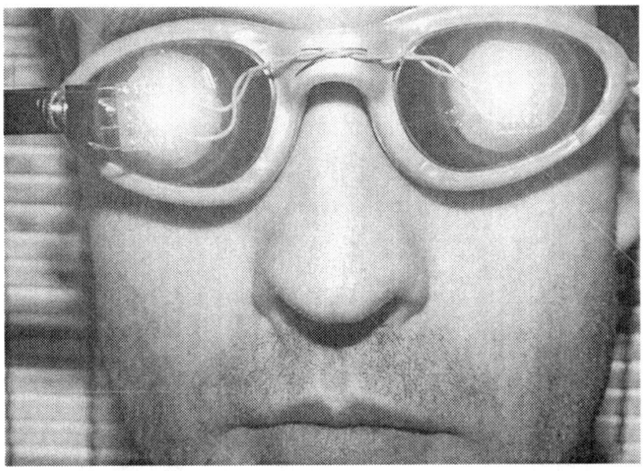

Figure 15-10 *"Jacked" into a more relaxing state.*

Project Sixteen

Clairvoyance Tester

Some of us know at least one person who claims to have some clairvoyance or an uncanny ability to predict future events. Being of the "believe nothing without proof" discipline, I decided to build a simple box that would allow my all-powerful psychic contacts to prove their worth or, as it seems more often, disprove their abilities! The word *clairvoyance* comes from seventeenth-century French and means *clair* = "clear" and *voyance* = "visibility." Those who are clairvoyant are said to possess some extrasensory ability to see future events or visions before they actually occur. To date, there has been no undisputable proof that anyone possesses this ability, and the practice is not normally accepted by the general scientific community.

This device allows your subject to use his or her amazing powers to "visualize" a number between 0 and 9, enter it on an LED readout, and then press a button to generate a completely random result on another LED readout. If the subject gets a match, then a third LED readout adds a 1 to the current score. If the subject's score is already zero, then he or she just stays at zero score. Since there is a 1 in 10 chance that the subject's guess will be correct, it is unlikely that his or her score will be anything other than zero or stay above a few points for any length of time. If your subject is truly clairvoyant, then surely this basic test will be easy for him or her, and the score should remain well above 3 after a few minutes of testing. Of course, this is almost impossible, and anyone who can score anything above 1 on a regular basis is a truly amazing psychic with unreal abilities.

The schematic for the clairvoyance tester, shown in Figure 16-1, does appear to have a lot of connections, and this is mainly due to the triple seven-segment LED displays used to show the guess, the result, and the score readouts. There are not enough pins on the small Atmega88 microcontroller to connect all three LED displays at once, so the three transistors (Q1, Q2, and Q3) switch between the three displays at such a fast rate that they all seem to be on at the same time. This time-sharing trick is how most LED displays work when there are hundreds or thousands of LEDs to control and only a limited number of connecting wires. There really is no limit to how many seven-segment LEDs you can connect as long as you have sufficient drive current and microprocessor speed and can spare the extra IO pin for each common connection. Other than the LED-display circuitry, the only other part of this circuit is the 4-MHz crystal resonator and the two pushbutton switches that allow the subject to select a number and then run the randomizer.

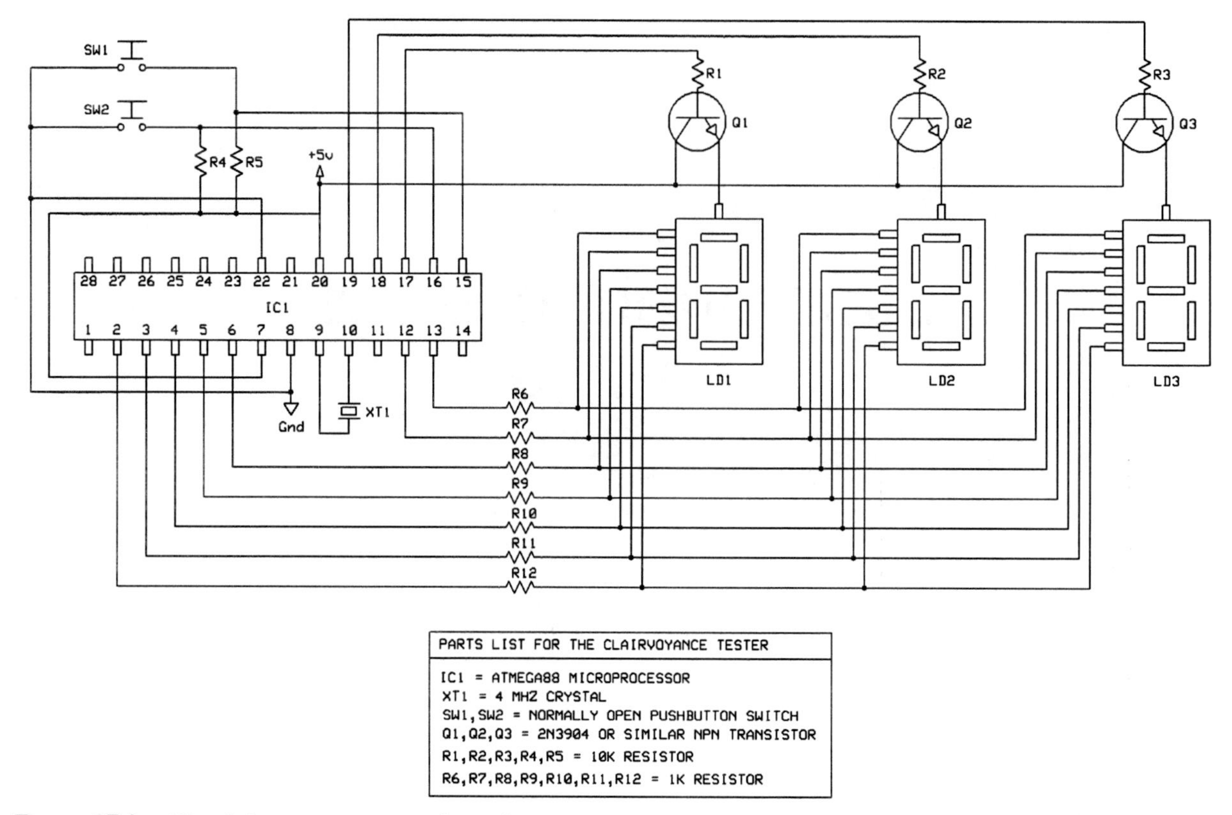

Figure 16-1 *The clairvoyance tester schematic.*

The seven-segment LED is a very common and inexpensive display that you are already familiar with because it is used in everything from your digital clock to the panel of your microwave oven. There are actually eight segments if you include the decimal point, but we don't use it in this application. These displays come as either stand-alone blocks or chained blocks containing more than a single digit. Some LED displays even have alphanumeric capabilities or multiple dots so that they can make any character imaginable. Figure 16-2 shows some of the LED displays that I found in my junk box when looking for three identical digits for this project. LED displays are either *common cathode* or *common anode*, which means that either the positive connections or the negative connections all go to a common point. It really does not matter which type you use as long as you install them in your circuit so that current is flowing in the proper direction. The ones I am using have part number HDSP-5501 and are basic green displays with a common anode, so the positive connections on all internal diodes are tied together.

To use a common-cathode LED display in this circuit, you would have to tie the driver transistor emitters to ground and connect the common cathode to the collector instead. Other than that, it is only for aesthetic purposes that all LED displays should be the same.

There are a lot more wires to connect in this project, so a breadboard prototype is important to ensure that the display shows the expected results, not some random hieroglyphics.

Figure 16-3 shows the schematic up and running with the triple-LED display reading "662," the results of over 2 minutes of clairvoyance testing. The first digit is my guess, and the second digit is the result of the random-number generator. The third digit is my score,

Figure 16-2 *Various segmented LED displays.*

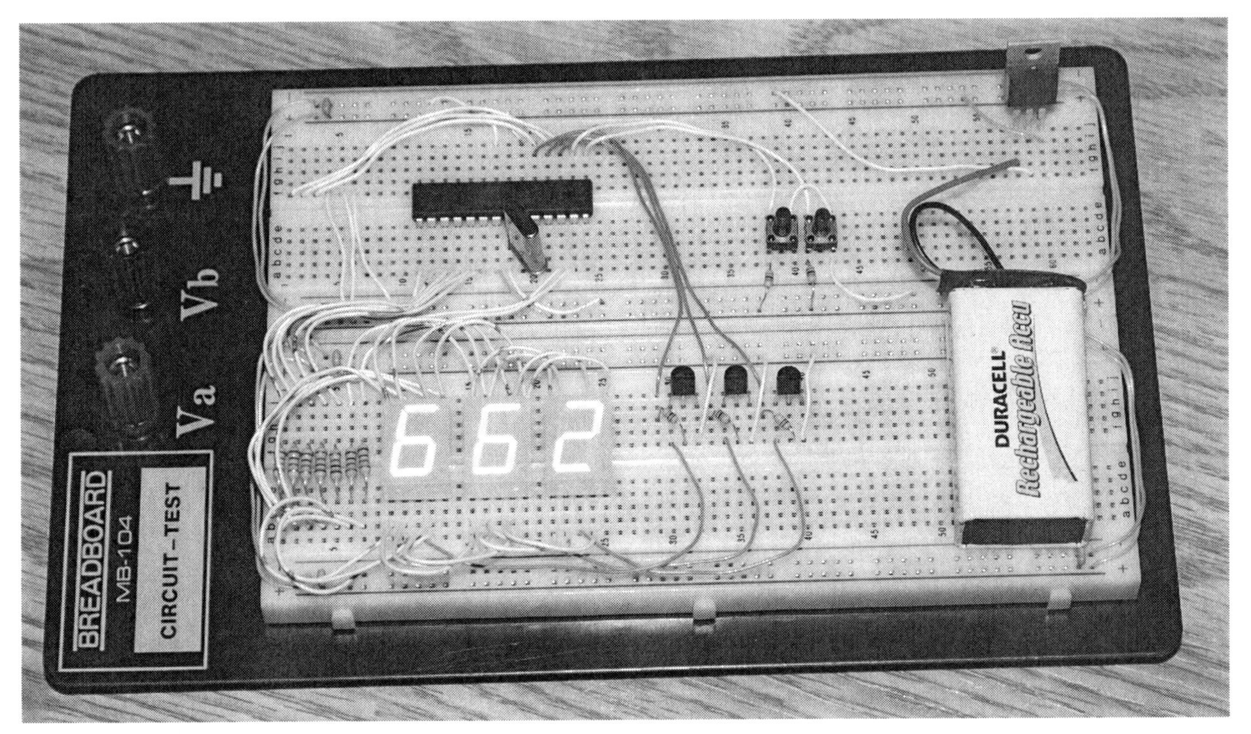

Figure 16-3 *Breadboarding the clairvoyance tester.*

which is usually 0 for the most part, so I thought I should grab a photograph of the unit. On the next two guesses, my score dropped back down to 0, so my belief that lotteries are a complete waste of time and money still holds strong. Notice the bank of seven current-limiting resistors to the left of the LED displays and the driver transistors to the right. Ninety percent of the wiring and hardware is just for the LED displays. I am also using a 7805 regulator to drop the 9-V battery down to 5 V, although the Atmega88 also would run fine from only 3 V supplied by a pair of AA batteries. As you lower the voltage, you may need to alter the value of the current-limiting resistors if your LED display is not bright enough. For 3-V operation, 500 Ω may be a better value than 1K.

If you have a bunch of LED displays and cannot locate a datasheet, then there is a simple way to figure out the connections to all pins. You will need a 5-V supply to your breadboard, a 1K resistor, and a few wires. Start by connecting the first pin to the resistor and then to VCC. Now take a wire from GND and try all other points to see if you get any segment to light up. If not, place the resistor to GND and the wire to VCC and then try again on all points. Keep moving along all the pins until you find the common pin and its polarity.

To keep the cabinet size to a minimum, I broke the circuit into two parts: the LED display perf board and the microcontroller perf board, as shown in Figure 16-4. Once all the wires were added, I was able to fold the two perf boards together like a stack to reduce the amount of space needed inside the small black box. The top of the box will have a square cut to fit the LED display by cutting over a line using a very sharp razor knife. A notching tool is also great for cutting out square holes for displays and other things in a plastic or thin-metal box. With a notching tool, you just drill a hole at each corner and then nibble your way across a line, taking one small bite at a time.

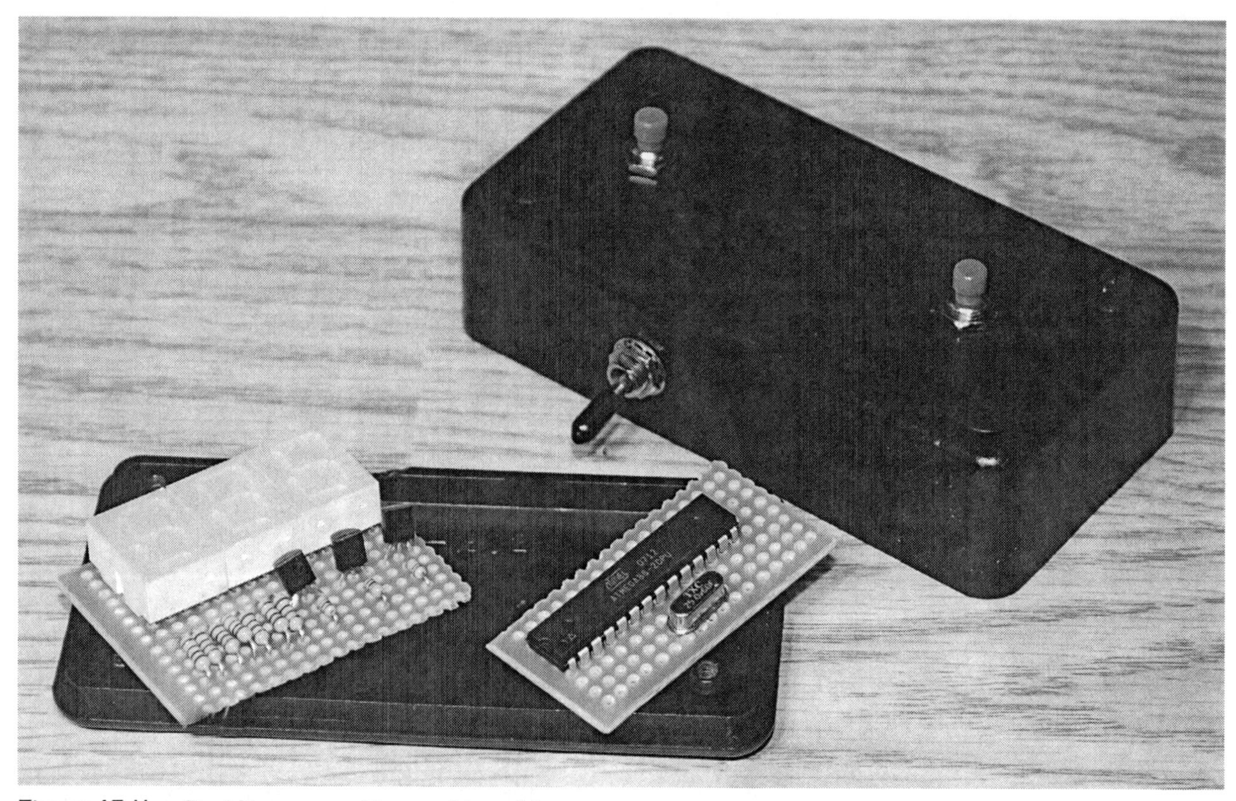

Figure 16-4 *Deciding on a cabinet and board layout.*

The completed clairvoyance tester is shown in Figure 16-5 displaying what is probably my highest score since building the unit. My original guess was a 7, and I just pressed the "randomize" button three times, and by chance, it resulted in a 7 each time. The odds of this happening are the same as matching any number three times in a row and quite rare. It would be the same as throwing a 10-sided die three times in a row and guessing the result each time before throwing. If a normal "nonclairvoyant" person were to use this device all day long, the score would remain at 0 most of the time, with the odd spike up to 3 or even 4 while "fluking out" once in a while. If you can find a person who keeps his or her score above 0 consistently, then please tell him or her to get ahold of me immediately because I will have a lot of work for him or her to do, such as choosing lottery numbers and reading the stock market. Anyone capable of beating the odds certainly will be living a life of luxury and fame!

Now let's look through the Basic source code that will be compiled into the Atmega88 microprocessor. There is also more information regarding microcontrollers and programming languages in Section Two with the project called "Lucid-Dream-Mask Controller" in case this is your first attempt at programming a microcontroller. I will step through each block of the program code, which is shown in Listing 16-1.

Have a read through the complete Basic source code of Listing 16-1 provided in the appendix so that you can get an idea of what the code is doing. If you are an experienced programmer, then this trivial code is probably something you could write in 15 minutes from scratch, but if you have never written a program in your life, not to worry because Basic is called that for a reason and is easy to understand. The lines in the program listing that start with an apostrophe are comments, and I will explain the code in blocks after each comment.

Figure 16-5 *Showing off my "impossible" score!*

```
' DEFINE TARGET = MEGA88 @ 4MHZ
$regfile = "M88def.dat"
$crystal = 4000000
```

The code following this comment is required to tell Bascom that we are going to target the Atmega88 device and that our clock will be an external crystal resonator running at 4 MHz. Telling the compiler your clock speed becomes important when using commands that deal with timing-sensitive routines such as serial transmission or analog-to-digital readings. Defining the device also helps the compiler to generate user errors that have to do with input/output (IO) pins. In this way, you can't accidentally try to toggle an IO pin that does not exist on the actual device. Critical timing is not an issue in this program, but speeds over 8 MHz may make it difficult to press the buttons without adding to the delays.

```
' CONFIGURE IO PORTS
Config Portd.0 = Output
Config Portd.1 = Output
Config Portd.2 = Output
Config Portd.3 = Output
Config Portd.4 = Output
Config Portd.6 = Output
Config Portd.7 = Output
Config Portb.3 = Output
Config Portb.4 = Output
Config Portb.5 = Output
Config Portb.1 = Input
Config Portb.2 = Input
```

This block of code sets up the pins that will connect to the LED display (outputs) and to the two pushbuttons (inputs). You can change these around any way you like, but the way they are now makes wiring a lot easier because there is some order to the IO pins.

```
' DEFINE VARIABLES
Dim ___rseed As Word
Dim Led(10) As Byte
Dim A As Byte
Dim B As Byte
Dim C As Byte
Dim D As Byte
Dim E As Byte
Dim F As Byte
```

Basic uses *variables*, which are letters or words used to hold values. I like to use single letters such as *A*, *B*, and *C* for simple programs such as this one, but when you are working on a large, complex program, the use of more descriptive variable names is recommended. "TIMER2" and "REDLED1," for example, would be descriptive variable names that make a lot more sense in a huge block of code. Variables are also defined as the number of bits they are to contain, so in our code, "A" and "B" are 16-bit "Words," which can contain a value between 0 and 65535. Variable "C" is an integer that can range in value from –32768 to +32767. Variables "D" and "E" will only contain values from 0 to 255, so they are bytes. Although you could just define all variables using larger data types, this is a waste of memory space and will slow down your code. Also notice the odd variable name "___rseed"; this is a reserved name and has special meaning to Bascom. When you place a value in "___rseed," you mix up the random-number generator, which is important when dealing with software-generated random numbers. By *seeding* the random-number generator, you ensure that the sequence will be random each time the program is run. Without this seeding, the same sequences of numbers would be generated each time you powered up the microcontroller. Although this random-number generation is still not purely random, it is certainly good enough for this application.

```
' DEFINE LED DIGITS
Led(1) = 8
Led(2) = 187
Led(3) = 97
Led(4) = 49
Led(5) = 178
Led(6) = 52
Led(7) = 4
Led(8) = 185
Led(9) = 0
Led(10) = 48
```

This block of code sets up an array of 10 values for the variable "LED." Although this may seem confusing at first, the values correspond to which of the seven segments will be lit to display a particular number. To make this a little more confusing, the value in parentheses is actually 1 higher than the represented numerical value, and to light a segment, we want a low bit, not a high bit. Thus the value of "Led(9) = 0" says that to display the decimal number "8," we want all bits to be off. This means that all segments light up, and an "8" will be displayed. It can be a bit of a chore computing these values by adding port bits, but after you have done it once for your segment, the hard work is completed.

```
' ********** MAIN LOOP **********
Main:
Do
```

Everything from here on will happen continually until the word "Loop" is reached, which causes program execution to start again where it first encountered the word "Do." This is called an *endless loop* because it never stops unless forced to by another command or an error.

```
' PRESSED RANDOMIZE BUTTON
If Pinb.1 = 0 Then
Gosub Roller
End If
```

Here, we simply check to see if the user has pressed the "Randomize" button, which starts the routine that flips through a bunch of numbers like a casino slot machine. Since the button inputs are tied to VCC through the 10K resistors, a value of 0 means that a button is depressed. If a button is pressed, the code will jump to the "Roller" subroutine, which is discussed later.

```
' CHANGE GUESS BUTTON
If Pinb.2 = 0 Then
A = A + 1
If A = 10 Then A = 0
While Pinb.2 = 0
Gosub Ledshow
Wend
End If
```

This code block checks to see if the user pressed the button that changes his or her "guess" number and then increments the variable "A," which is used to hold the number. If the value of "A" goes beyond 9, then it is reset to 0 because the display can only draw single digits. Since microcontrollers are working in millions of instructions per second, this routine also does something called switch debouncing so that the microsecond state between switch contact and noncontact is not read as multiple presses. To debounce the switch, the code simply calls the "Ledshow" routine as long as the user is still holding the button down, which actually creates a bit of a delay and avoids false contact readings. Anyone working with microcontrollers and switches is definitely familiar with switch debouncing.

```
' RANDOM SEED COUNTER
___rseed = ___rseed + 1
```

Here, the random-number generator gets another seed number, which is just a rolling counter, made of the word variable "___rseed." This ensures that there is no way you will ever learn to read a pattern in the random-number generator so that no one can trick this device.

```
' DISPLAY DATA ON LEDS
Gosub Ledshow
```

Since the three LED displays need to be refreshed one at a time on the shared bus, this routine is called here each time the program loops, which is indefinitely. The "Ledshow" routine (discussed soon) ensures that something is always shown in the three LED displays.

```
' RESTART MAIN LOOP
Loop
End
```

The "Loop" command causes program execution to jump back to the main routine where the command "Do" was encountered. This is the end of the infinite loop, and all other commands beyond here must be called by either "Goto" or "Gosub" commands.

```
' ****** LED DISPLAY ROUTINE ******
Ledshow:
Portb.5 = 1
Portb.4 = 0
Portb.3 = 0
D = A + 1
Portd = Led(d)
Waitms 2
Portb.5 = 0
Portb.4 = 1
Portb.3 = 0
D = B + 1
Portd = Led(d)
Waitms 2
Portb.5 = 0
Portb.4 = 0
Portb.3 = 1
D = C + 1
Portd = Led(d)
Waitms 2
Return
```

This is the routine that displays a digit on each of the three LED displays. A little trick called *persistence of vision* is used here to switch between displays so fast that your eyes think that they are all on at the same time. As you can see, only one bit of the three "Portb" pins is on at a single time, and then the variables "A," "B," and "C" are sent to the display for only 2 ms each. Because 2 ms is so fast, it appears that each display is on all the time, and the seven-segment lines can be shared, saving valuable IO overhead. Since Bascom array variables start at 1, not 0, the line "D = A + 1" adds one to the array pointer so that the value in "Led(d)" is the same as the decimal value we want. This conversion just makes it easier to understand the code, especially when trying to compute the segment bits from scratch the first time. Once all three displays have been lighted for 2 ms each, this routine just "returns" to where it was originally called.

```
' ***** RANDOM ROLL ROUTINE *****
Roller:
' RANDOMIZE RESULT
For E = 1 To 50
B = Rnd(10)
For F = 1 To 10
Gosub Ledshow
Next
Next
```

This is the routine that starts the slot-machine-style random-number generator that will display the random digit on the second LED display. The line "For E = 1 To 50" indicates that the random number will "roll" past 50 times, making a more interesting display, as if the device were actually rolling a die or spinning the mechanics in a slot machine. Each time, the "Ledshow" routine is called another 10 times just to slow the display of random numbers down enough to actually see them go by.

```
' ADJUST SCORE
If A = B Then
C = C + 1
Else
C = C - 1
End If
If C = 255 Then C = 0
Return
```

After 50 random numbers flip past on the second LED display, this next block of code then checks to see if the "guess" displayed on the first LED matches the latest random number displayed on the second LED. If there is a match ("If A = B"), the score variable "C" is incremented by one. If there is no match and the score is above zero, it is decremented by one. The line "If C = 255 Then C = 0" checks to make sure that the score variable never goes below zero because a value of 255 means that the byte variable has wrapped.

So this is how the Basic code works. A lot can be accomplished in a very few lines using Basic, and since microcontrollers work at nanosecond speeds, you have a lot of power at your fingertips.

Now you can weed out the false psychics from those who possess real clairvoyant powers! It has been suggested that under the effect of the Ganzfeld device presented earlier, extrasensory perception (ESP) and clairvoyance may increase for certain people, so this might be an interesting experiment to try. Have your subject guess a number, and then enter it on the clairvoyance tester for him or her, each time letting the machine keep score while the subject meditates. Maybe you want to train your ESP so that you can get the upper hand next time you play a dice game? All you have to do is alter the code slightly so that it displays only the digits 1 to 6, and your clairvoyance tester is now a dice-roll simulator. It even would make a great decision maker or electronic dice roller for game night!

Have fun weeding out your false wizards and psychics with this project. If you do find someone to be especially clairvoyant, then please tell them to get in contact with me for a job interview immediately. My phone number is . . . wait, let them guess it! Next, we will try our hand at hypnosis.

Project Seventeen

Visual Hypnosis Aid

When I began to research hypnosis in order to come up with a nice single-paragraph description of what it is all about, I discovered that there is currently a huge amount of controversy as to what the definition of hypnosis really is. Some people say that it is a trancelike sleep, whereas others say the exact opposite, claiming that it is more of a heightened conscious state. So rather than wave the flag and take sides, I will just concentrate on the methods and devices used to help a hypnotist bring his or her subject under hypnosis and present an easy-to-build visual hypnosis aid that you can use to learn the art yourself.

Obviously, hypnosis is a very powerful tool that can be used to help us with our inner problems, break us from long-time habits, delve into the subconscious, and unlock the mind in ways not possible without it. Hypnosis, like all skills, can be learned by anyone willing to read the literature, and you can even practice on yourself, exploring self-hypnosis. Of course, it is highly unlikely that you will have your friends under your complete control by speaking the "secret phrase," nor will you learn to mass-hypnotize a group of people into running around barking like dogs because this is the stuff of movies and rehearsed magic!

This device will replace the classic pendulum, or "swinging watch," that is so often shown in movie scenes involving hypnosis. Often taken to the extreme, the patient simply will fall into a trance as the doctor lets his or her watch swing back and forth in front of the patient's eyes. This scenario is highly unlikely, but the use of a fixation point to keep the mind focused during hypnosis is a viable method of induction. Instead of a swinging object, this device simulates a back-and-forth motion using six bright LEDs, allowing it to be used in a low-light room or in the dark.

As shown in the schematic (Figure 17-1), the six high-brightness LEDs are actually connected to the 10 output lines of the 74HC4017 decade counter (IC2). Since the goal is a back-and-forth motion much like a swinging pendulum, the sequence counts as follows: 1, 2, 3, 4, 5, 6, 5, 4, 3, 2 and then repeats, which is why we need 10 sequential lines and six LEDs. The 4017 is clocked by the 555 timer (IC1), which can have its rate controlled by adjusting the variable resistor (VR1). The range of speed is quite large, so you can create a nice, slow back-and-forth light sequence or have the LEDs moving so fast that they almost seem to flicker. The 10 diodes connected to the outputs of the 4017 block

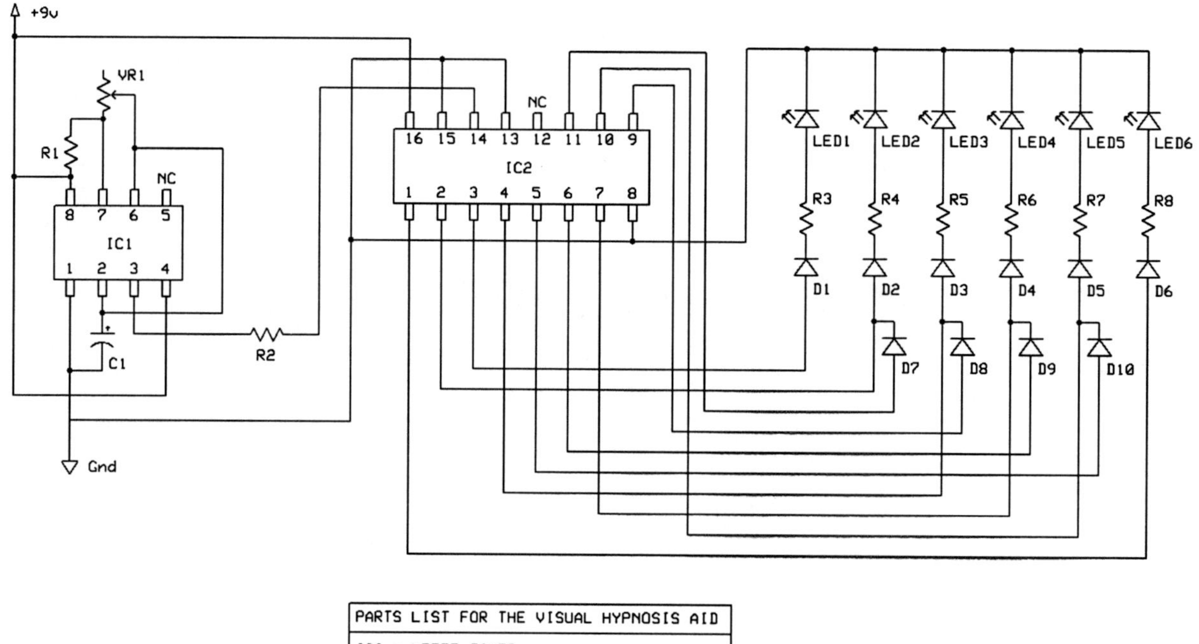

PARTS LIST FOR THE VISUAL HYPNOSIS AID

IC1 = NE555 TIMER
IC2 = 74HC4017 DECADE COUNTER
C1 = 1UF
VR1 = 10K VARIABLE RESISTOR
R1,R2 = 1K
R3,R4,R5,R6,R7,R8 = 1K
D5 - D10 = 1N4003 OR SIMILAR DIODE
LED1 - LED6 = HIGH BRIGHTNESS LEDS

Figure 17-1 *Visual hypnosis aid schematic.*

current from feeding back into the output pins when counting back over the last four digits to reverse the sequence. The resistors connected in series with the LEDs are there to limit the current draw and can be replaced with a lower value if you want your LEDs to give off more light. There is a limit to how much current the 4017 can source and how much your LEDs are going to tolerate, so the values presented here are a good start. If, for some reason, you want extremely bright LEDs, you will have to drive them with a transistor rather than taking current directly from the outputs of the decade counter.

The visual hypnosis aid device is shown working on the solderless breadboard in Figure 17-2. The circuit is quite simple to build and tolerant of supply voltages from 3 to 12 V, although you may have to adjust the current-limiting resistors to set the optimal brightness for your LEDs. The high-brightness type of LEDs work best for this device, and they will allow you to project the light onto a wall or screen for easier viewing, as will be shown later. Color is totally up to you, and there are many ultrabright LEDs available in all colors and white. A single 9-V battery is used to supply power because this was the most convenient setup.

The circuit was transferred from the breadboard to a small perf board, as shown in Figure 17-3, after deciding on what type of LEDs would work best. This LED sequencer is a cool gadget that can be used for all kinds of display purposes, such as electronic props, computer case mods, or anything that would look better with a little light show. If you want to use more that six LEDs, you can double them up in parallel, but in the end, you still will only have 10 steps and may end up needing a driver transistor for each step to handle the extra current.

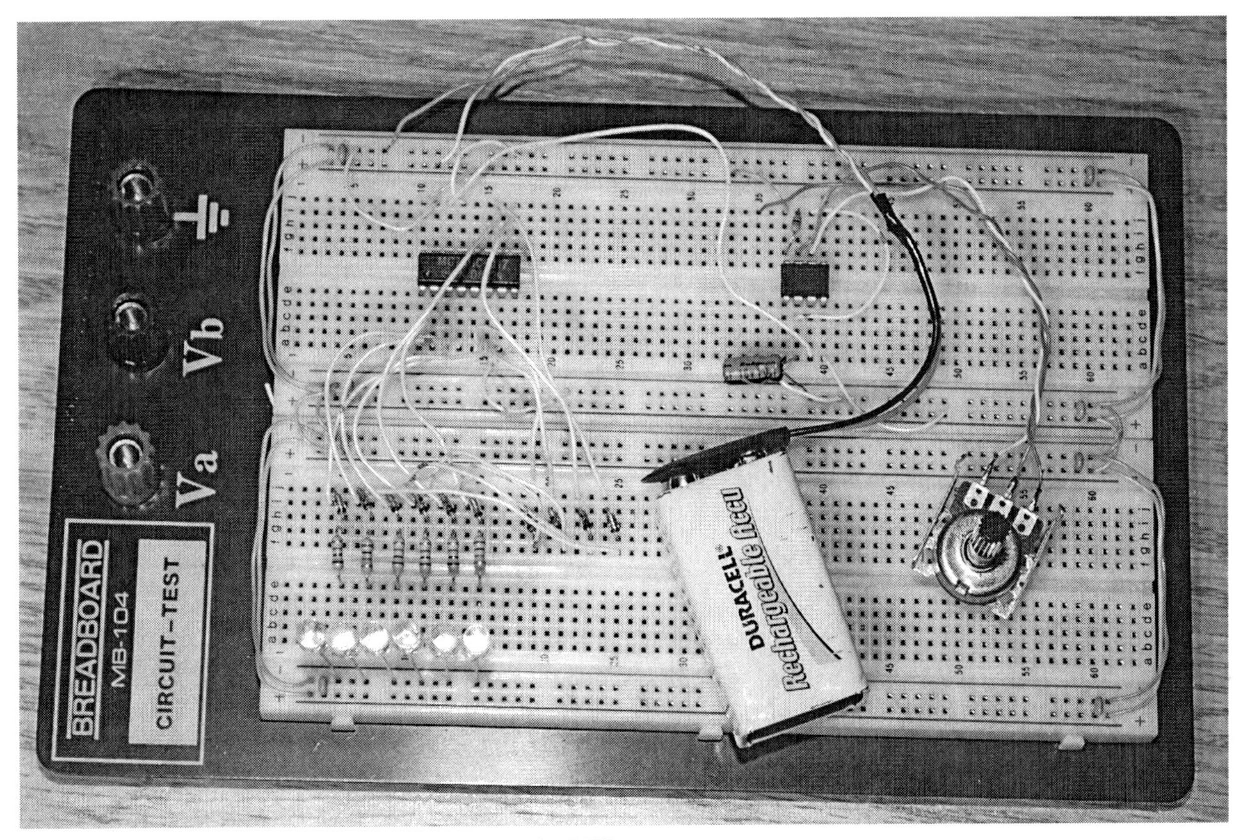

Figure 17-2 *Breadboarding the circuit to test the LEDs.*

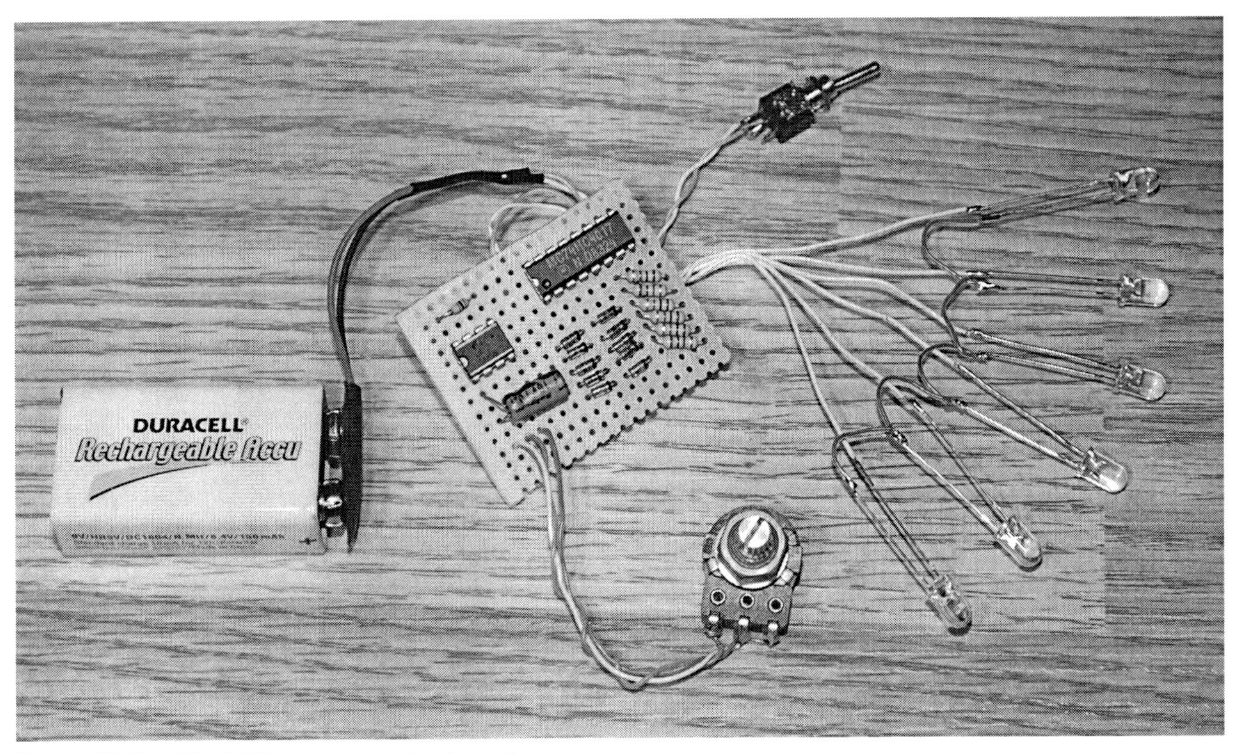

Figure 17-3 *The LED sequencer circuit ready to run.*

There are numerous configurations this device can take, and it really depends on how you want to deliver the light to your subject's eyes. The distance between each LED and the focal range of the LEDs will affect how far away the unit can be from your subject. I found that a spacing of no more than 1 in between the LEDs was fine, and anything beyond that began to ruin the smooth back-and-forth effect of the unit. To place the LEDs at a greater distance apart so that the device will work when it is placed farther from your subject, you will need some type of diffusing lens to spread out and smooth the beam from the high-brightness LEDs. Figure 17-4 shows the all-in-one configuration I used for my visual hypnosis aid. The spacing between LEDs is about 1 in.

Although the single boxed unit worked well in a dark room if placed about 2 ft from my eyes, I wanted the ability to experiment on more than one subject at a time, so I came up with a simple method that would expand the size of the display by projecting the light onto a diffuser screen. This setup allowed viewing of the hypnosis aid from a much greater distance and by many people at the same time. Depending on the brightness and focal length of your LEDs, you may be able to project them several inches or even several feet. Try aiming your device at a light-colored wall in a dark room to see what you get. My LEDs had the ability to project their light on a wall almost 4 ft away, so my goal of a 10-in screen certainly was within reach. Figure 17-5 shows how I made a basic frame for a white-paper screen by bending a coat hanger with a pair of pliers.

To project the light evenly to the screen in front of the LEDs, I had to drill the mounting holes a bit larger and then angle them slightly in opposite directions, starting from the center two LEDs. A hot-glue gun was used to hold the LEDs in place so that they would beam outward, as shown in Figure 17-6, while projecting to the paper diffuser screen. I also mounted the cabinet on a tripod so that it could be positioned easily in a room at any height and angle for easy viewing.

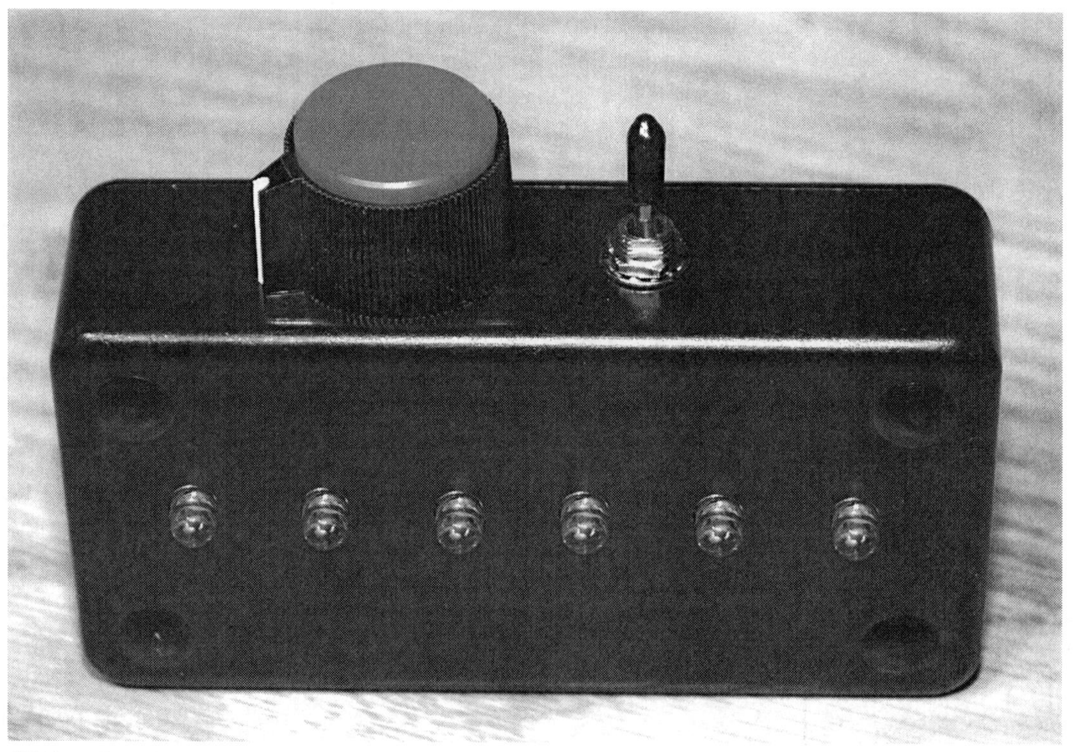

Figure 17-4 *Completed visual hypnosis aid unit.*

Figure 17-5 *A simple frame for a diffuser screen.*

Figure 17-6 *Diffuser screen lit by the LEDs.*

Now all you have to do is a little reading on hypnosis techniques and find someone willing to fall under your newfound powers. Hypnosis is really all about trust and belief, so don't expect much of a result from those who think they cannot by hypnotized because chances are they can't. You can always dig into self-hypnosis and never have to look for a willing participant! Some experiments you might want to try include setting the sequencer to a frequency in the alpha-wave band, changing the LED colors, or even taking a line out of one of the 4017 pins and feeding an external device to produce synchronized sound or more light. Have fun with this device, and if you manage to acquire hypnotic powers like Reveen, remember to be nice to your "zombies"!

Project Eighteen

Color-Therapy Device

Color therapy, also known as *chromatherapy* and *colorology*, is the process of balancing one's body and mind using color and light. As with many alternative medicines, color therapy is believed to be viable by some and is cast into the pit of pseudoscience by others. Color therapy has very old roots that go back thousands of years to the ancient Egyptians, who used giant solariums and colored-glass filter lenses to change the color of sunlight. On a more modern note, in the early 1920s, Swiss psychotherapist Max Luscher developed a well-known test using color that measures a person's psychophysical state based on the subject's color choices.

The power for color to evoke certain moods and responses is certainly undeniable, and advertisers have known this for decades. Have you ever noticed how many restaurant chains choose orange and yellow? How about financial institutes or other organizations that want your trust? They prefer to draw their presence in blue or magenta. There are reasons for many of the common color choices. The following is just a small sampling of colors and their accepted meanings.

Red says, "Look out!" Obviously, it screams out danger and commands attention. War, strength, love, desire, and passion are just some of the emotions that red represents. For this reason, red is the perfect color for warning signs, emergency vehicles, or anything else that needs to command attention. Did you know that people who own red sports cars pay more for their insurance? They are considered thrill seekers because of their color choices.

Orange is a happy color. Being a combination of red and yellow, it stands out, but in a way that does not say, "Look out!" Happiness, stimulation, and contentment are a few of the emotions associated with orange. Orange is the perfect color for a fast-food chain because it implies instant happiness. Orange also has been known to evoke feelings of hunger because it is also associated with citrus and good food.

Yellow is the color of the sun, so it has a warming effect and evokes feelings of happiness and joy. Energy, well-being, cheerfulness, and intellect are some concepts that work well with the color yellow.

Blue is the color of the sky and often is associated with trust, stability, loyalty, confidence, and strength. Blue is used to represent financial institutions or any organization that wants to convey its trustworthiness. Blue is also related to tranquility, peace, and serenity. Blue is also considered a masculine color, especially the darker shades. Blue also has the opposite effect that orange does when it comes to our appetites, so there are not many fast-food chains using blue as the main color in their logos.

There are so many more examples of the link between color and emotion that they probably would double the size of this book. The real question is, Can color be used as a tool to heal the mind and change our mental state?

Using the simple device presented here, you can make your own decisions about color therapy by performing your own experimentation. Using only two LEDs and a few variable resistors, you will be able to generate any color that nature can generate in a controlled fashion.

An RGB LED is really just three separate LEDs in one small package. By placing the tiny light-emitting chips on one small base, you actually can mix the three colors together to produce any possible color because an RGB LED contains a red, green, and blue LED. By adding a variable resistor to each color, you can vary the intensity of each primary color and create millions of different colors exactly the same way your computer monitor and television can. You can use an actual RGB LED or just find a set of red, green, and blue LEDs with the same style of lens and use them to mix colors. Figure 18-1 shows some of the possibilities I found in my junk box, along with a $2 color candle that actually contained an RGB LED inside. The discount-store color candle actually was the same price as the bare RGB LED and even came with a battery, so it pays to scrounge around the dollar stores for parts once and in a while.

An RGB LED looks like any ordinary LED except that it will have four leads, not just two. There will be a common cathode or anode as well as the three separate color leads. LEDs that have three leads are only two-color LEDS. The inside of the dollar-store candle is shown in Figure 18-2, which reveals a typical clear-lens RGB LED that is often sold for more than a dollar at an electronics supply store. If you are shopping around for new RGB LEDS, try to find one with an opaque casing rather than the clear type because the clear LEDs need some kind of

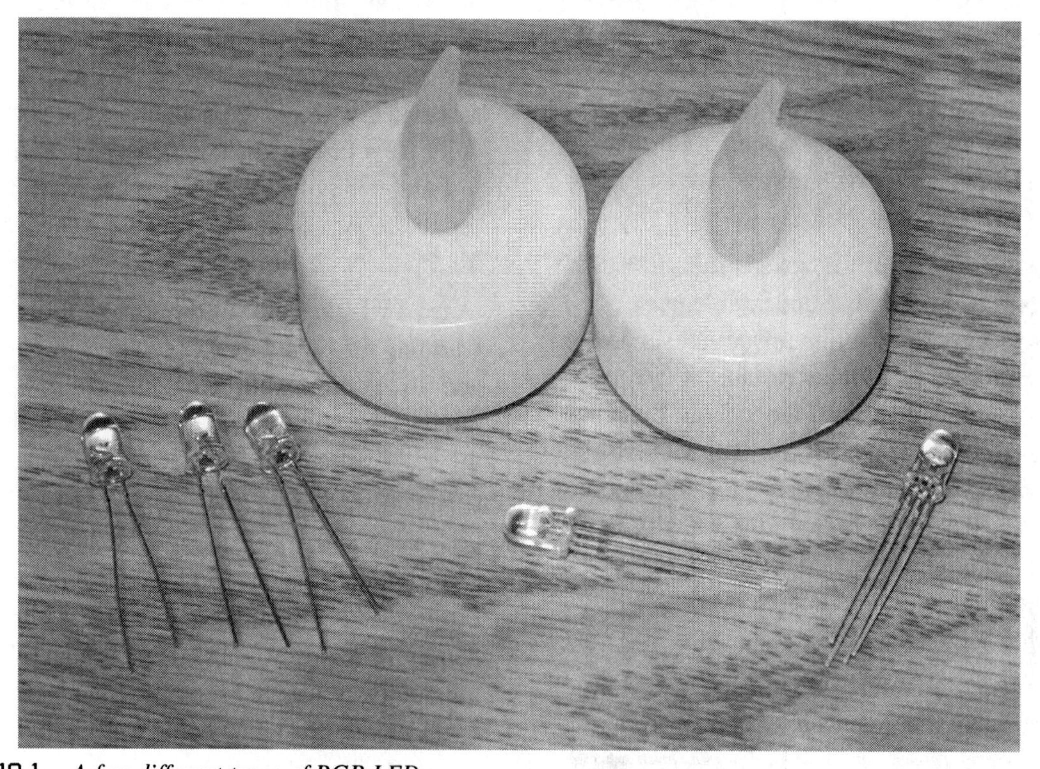

Figure 18-1 *A few different types of RGB LEDs.*

Figure 18-2 *Inside the dollar-store candle.*

diffuser to better mix the colors. Since there are actually three separate LEDs inside an RGB LED, there will be three distinct beams emitted when all the LEDs are lighted. Diffusing the three LEDs makes the color mixing look much more convincing.

Another quirky thing about RGB LEDs is that each color will have completely different current and voltage requirements. Often the red LED will have a much lower rated voltage and be made of a different material than the green and blue LEDs. Since I decided to use the new store-bought RGB LEDs (bottom right of Figure 18-1), I had a proper datasheet, which helped when choosing the current-limiting resistors. In my device, the red LED was rated for a maximum of 2.5 V, whereas the green and blue LEDs had a maximum rating of 3.5 V. Testing this with a single 1.5-V battery, it was obvious that the red LED was much too bright at the same voltage as the green and blue LEDs. For this reason, the current-limiting resistor on the red LED shown in the schematic (Figure 18-3) is three times the value of its green and blue counterparts. By tuning the current-limiting resistors to each LED, you then can control the brightness using three identical variable resistors so that they operate in a linear fashion when making up the endless color possibilities. If all variable resistors are set to the same rotation, you will have some value of white. Changing them to random positions will create every color possible.

The three current-limiting resistors shown in the schematic may not be the best values for your RGB LED, so you may need to experiment a bit to get white out of your LEDs with all variable resistors turned up full. To play it safe, start with 1K current-limiting resistors, and then look at the datasheet for your LED. Red likely will have a much lower operating voltage, so you can add resistors in series like I did until all three colors seem to be the same brightness level (Figure 18-4).

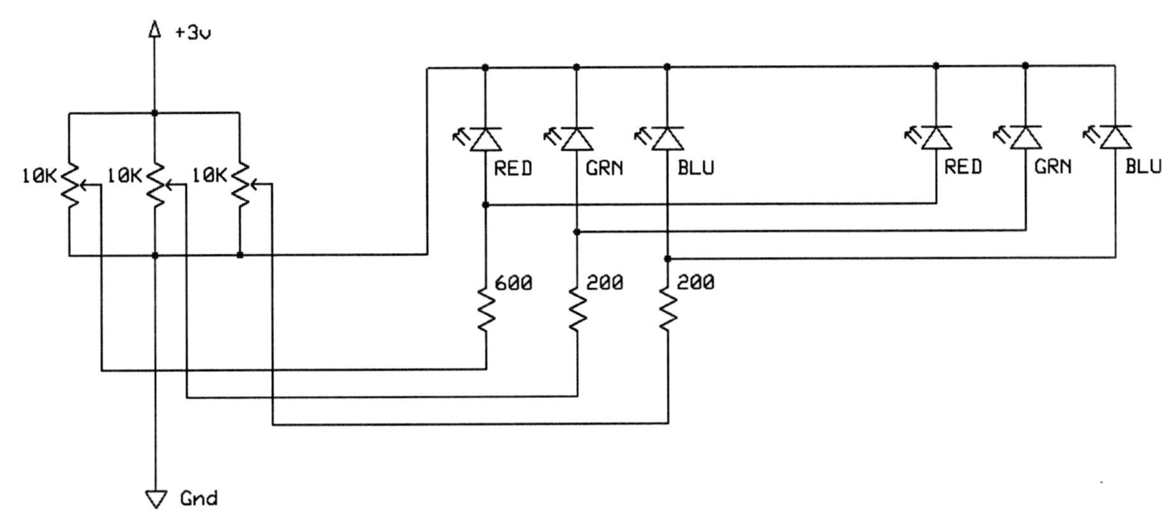

Figure 18-3 *Connecting the LEDs to a single power source.*

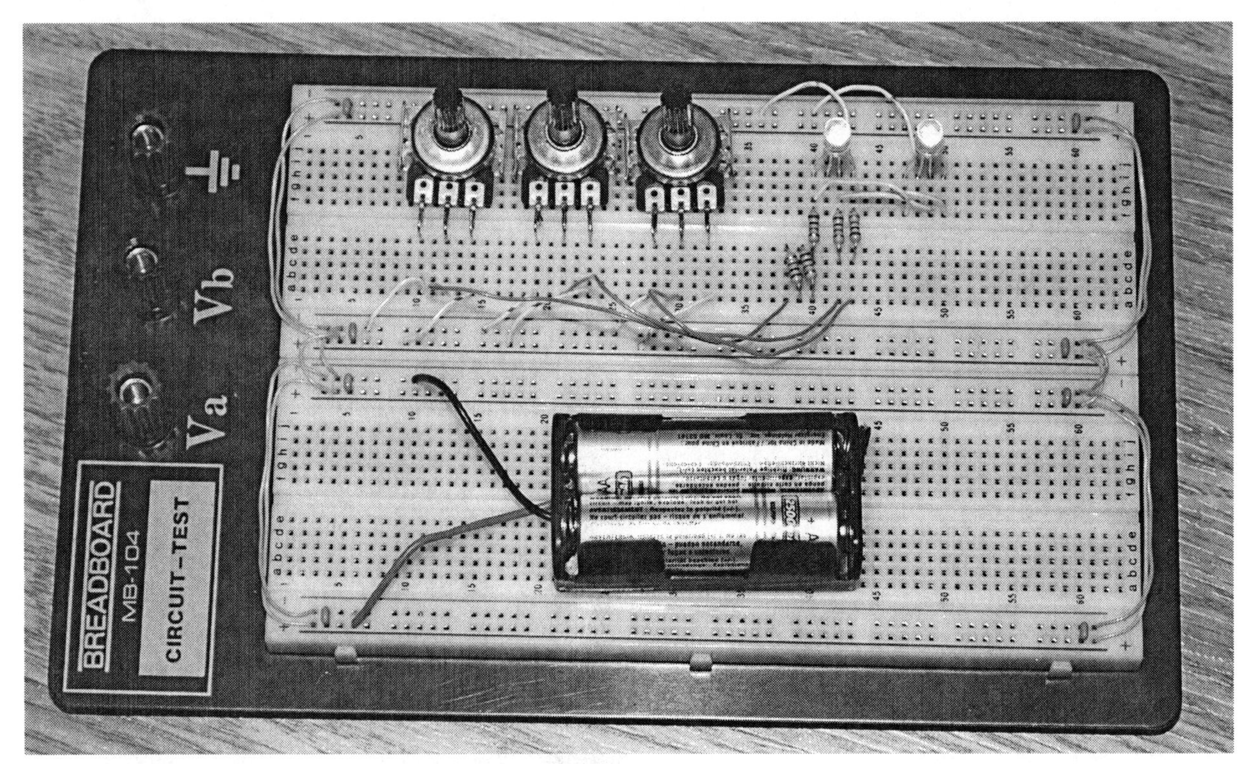

Figure 18-4 *Testing the brightness of the LEDs on a breadboard.*

Place a piece of paper over the LED to create a simple diffusion screen when checking the levels so that you can see a better color mix. If all colors are outputting the same level of brightness, you should see white.

Figure 18-5 shows the three small LEDs contained under the clear LED casing, which is why three separate beams will be emitted from the device. I am not sure why they are made like this because it really defeats the entire purpose of having an RGB LED, but there are ways we can solve the problem and get a much better color mix. A simple way to diffuse your LED is to just sand down the casing until it is so rough that you can't see through it anymore. If you are lucky enough to find an RGB LED in a milky white casing, then you will not have this problem, although these seem hard to find for some strange reason. The good news is that you can make a perfect diffuser for your eyes using some Ping-Pong balls, as was done earlier in this section.

By placing half of a Ping-Pong ball or a piece of white Styrofoam over the LED, you can really diffuse the three beams a lot better. Without the diffusing screen, it is almost impossible to see that the colors are mixing, but by placing the Ping-Pong ball half over the LED (Figure 18-6), a much better mixing of colors can be had. When the camera took the photograph, it was still easy to see the three separate beams, although the color mixing was quite good when viewed by eye. For even more diffusion, some white foam or paper slices can be placed inside the Ping-Pong ball half to further scatter the light, although this will cost you in brightness.

If you plan to deliver the LED light to your eyes using a mask or some type of goggles, then a perfect diffuser tube can be made out of a pair of film canisters and a pair of cut-in-half Ping-Pong balls. Place the Ping-Pong ball into the open end of the film canister, as shown in Figure 18-7, so that the nonlogo side will be used. Trace around the top of the canister so that you can cut out a segment of the ball that will fit perfectly back into

Figure 18-5 *You can see the three small LEDs inside.*

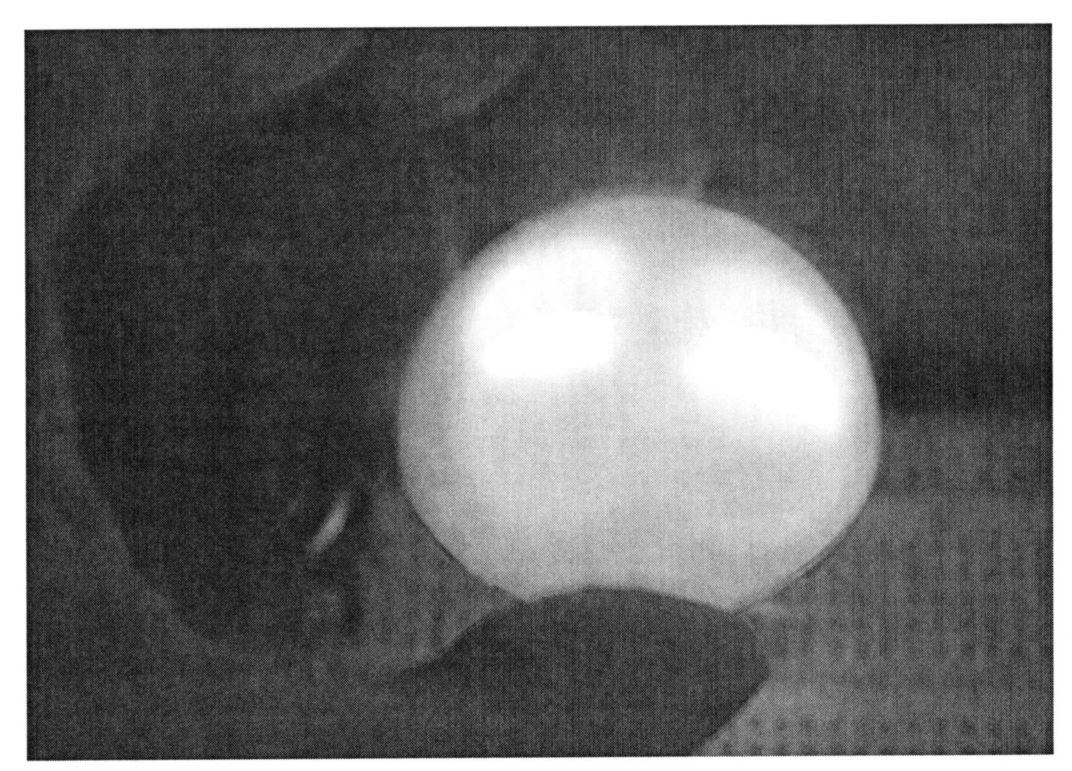

Figure 18-6 *Better diffusion of the three colored beams.*

Figure 18-7 *Making a diffuser tube for the goggles.*

the canister. A pair of scissors with a sharp tip is best for cutting the balls to avoid cracking the brittle shell as you cut along your line.

The two Ping-Pong ball segments are affixed to the open ends of the film canisters by hot gluing them around the edges, as shown in Figure 18-8. If you find that your LED works best with a bit of white Styrofoam in front of the ball half, then don't forget to stuff it into the canister before you glue in the ball segments.

Figure 18-8 *Gluing the ball halves to the canisters.*

Since you won't be wearing the goggles for extended periods, a pair of inexpensive swim goggles will work just fine for this project once you loosen the straps so that they are not too tight. If you can't find clear plastic goggles, then you will have to use a Dremel tool or soldering iron to cut or melt out the front lens so that the color of the light is not affected. My lenses were clear and I also painted the inside using a black marker to block out any ambient light that may enter from the sides. Figure 18-9 shows the diffuser tubes glued to the front of the goggles, as well as the RGB LED stuck into a hole at the other end. This configuration worked very well because the length of the film canister was almost perfect for the focal length of the LEDs. The three beams came very close together over the Ping-Pong ball surface, which made for a good mixing of colors.

The RGB LEDs also were hot glued to the film canisters once the leads were trimmed and the wires installed. If you plan to use the goggles on another person or with some other type of LED driver, then you also might consider using a removable plug to connect the goggles to the control box. I just used a 6-ft length of four-

Figure 18-9 *Diffuser tubes glued to the goggles.*

conductor wire to connect between the box and the goggles because I was only using the color-therapy device on myself. Eventually, you run out of victims, I mean subjects, as your friends get to know your strange "mad scientist" side!

The control box is just your three variable resistors, current-limiting resistors, and battery pack placed in a box, as shown in Figure 18-11. Although you probably don't have to label the controls, it does help when you are not the subject and are experimenting with your subject's

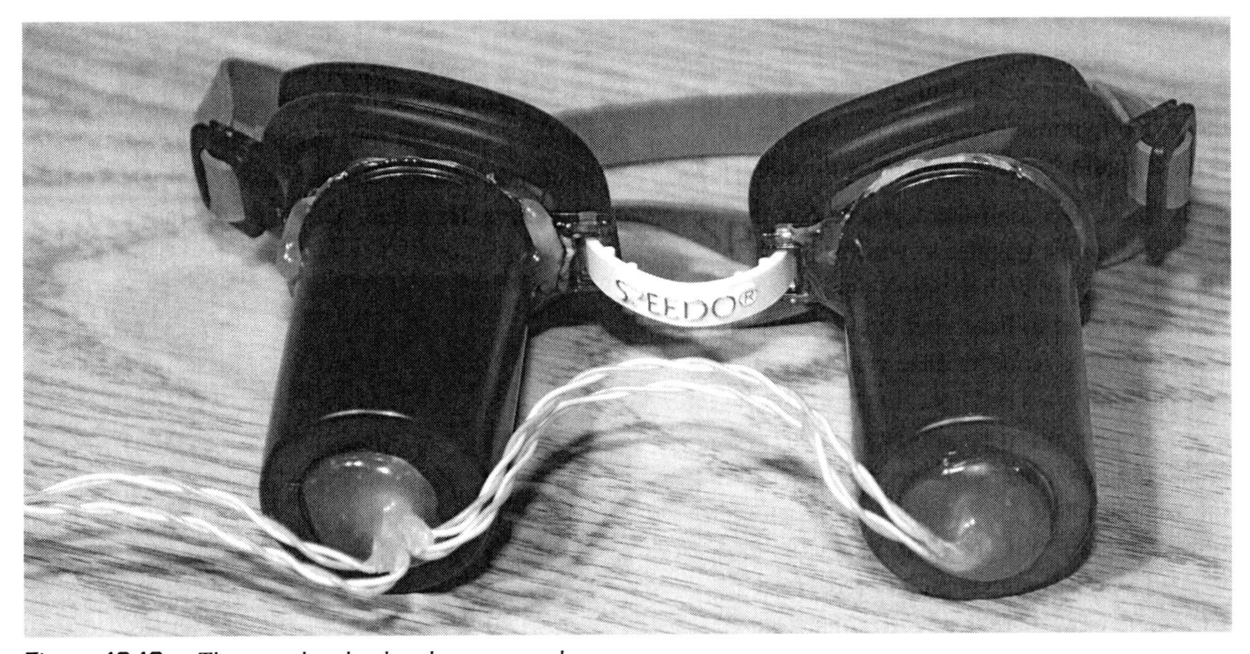

Figure 18-10 *The completed color-therapy goggles.*

Figure 18-11 *The completed color-therapy control box.*

reactions to the various colors. Since the circuit will slowly drain the batteries even if the LEDs are turned right down, an on/off switch will be necessary. With a pair of AA batteries, this device will run for many days because the LEDs use very little current. Another modification you might want to consider is an input on the side of the control box to connect some external device to the LEDs. Some of the devices presented earlier, such as the alpha meditation goggles and the visual hypnosis aid, could be modified to drive the color-therapy goggles with little work.

Operating the color-therapy device is easy. Simply strap the goggles to your head (Figure 18-12), and play with the dials until you feel great. You need to have your eyes open for the color device to work because your eyelids are red and will only pass red light. I actually found that a few minutes of bright yellow each day did wonders to help beat those winter blues I often get when it's –40°F outside and dark before suppertime. Even if you don't believe in the whole balanced mind and body aspect of chromatherapy, there is no doubt that colored light will affect your moods, and this device has some real merits. I tried the orange-light experiment to see if it would affect my hunger, but I cannot honestly say if it worked because the thought of eating was already in my mind when I dialed up the orange color.

For a nonbiased experiment on the various colors and their known responses, try the device on a willing subject who has no clue about which colors are for which emotions. If you search for "color therapy" or "chromatherapy" on the Internet, you will find many sites that link a multitude of colors to various emotional responses. Have fun, and enjoy the rainbow!

Figure 18-12 *Ah, yellow . . . a cure for the winter blues!*

Project Nineteen

Synchro Brain Machine

This interesting device is based on research done by Professor Heinrich Wilhelm Dove in 1839. Dove discovered an effect called *binaural beats* in which the brain will hear a "beat" when presented with two slightly different independent audio frequencies into each ear. This effect is similar to what is heard when one tunes up an instrument while using another sound source as a reference. That unmistakable "wong, wong, wong, wong" sound as a guitar player tunes up a string is another example of the "beat" that can be heard as two sound waves go in and out of phase with each other. In the brain, something more interesting happens if we are fed the two slightly differing frequencies into each ear. It is called *brain-wave entrainment.*

Brain-wave entrainment also was the focus of a previous project in this chapter, the alpha meditation goggles, although it is said that by using binaural beats, the brain entrainment effect is much stronger. The general theory is that the resulting frequency of the binaural beat will coax the two brain hemispheres to become synchronized over time. By setting the oscillators so that their phase difference is somewhere between 10 and 15 Hz, your brain will subconsciously hear the alpha frequency and eventually fall into sync with it. The actual frequency of the oscillators is much higher than the resulting binaural beat, often in the range of hundreds of hertz.

For example, if you wanted to entrain your brain to the frequency of 10 Hz, you could set the right-ear frequency to 315 Hz and the left-ear frequency to 325 Hz, resulting in a binaural-beat frequency of 10 Hz. The various frequencies of the brain are named delta, theta, alpha, beta, and gamma. They have the frequencies and effects on consciousness as follows.

Delta waves are frequencies less than 4 Hz. This means that there are only four cycles per second, so this is a very low frequency. When the mind is in the deepest stage of sleep, where there are no dreams and the body is fully paralyzed, delta waves are prominent. It is probably not much use trying to entrain your brain to this frequency because you would have to traverse the other states of consciousness to reach this level.

Theta waves occur between 4 and 7 Hz, overlapping into the lower end of alpha waves. When your brain is running at the theta frequency, you are likely either dreaming or in a state of very deep mediation. If you entrain your brain to the theta state, the result likely will be that you fall asleep.

Alpha waves fall between 7 and 15 Hz. Entraining the brain to the alpha frequency is usually the goal of entrainment because it provides a person with pure relaxation without falling asleep. At the lower end of the alpha scale, a person is just starting to cross the line between deep relaxation and sleep, so this is an interesting frequency to experiment with. At the higher end of the alpha range, a person is fully conscious but feeling very relaxed, so reaching this state a few times a day can be a real benefit to the mind.

Beta waves occur between 15 and 40 Hz. This is the state of mind we are in while working or thinking. Your brain probably is working on some part of the beta frequency range right now as you read this book, and it is the most common state of mind during the day. Trying to entrain the brain to this frequency might not be much use because you are probably already working on beta waves if you are getting ideas about doing these experiments!

Gamma waves are the frequencies that occur above 40 Hz. If your brain is in the gamma state, then you are really working your gray matter because this state of mind includes fear, difficult problem solving, anxiety, and many other higher mental activities. You might want to entrain your brain to the gamma frequency when trying to work out a very complex problem or right before some type of test that will push your mental abilities to their limits. The gamma state is certainly not a relaxing state, but it does have its uses.

One extremely interesting and rare state of mind is called *hypnagogic sleep paralysis*, which is an almost accidental state of sleep and consciousness at the same time. If you have ever been through a sleep-paralysis episode, then you know what I mean when I say the experience can be terrifying, especially if you have no idea what is going on. You open your eyes, but your body fails to respond, although you have complete control over your eyes. Many people experience sleep paralysis a few times during their lives, often just after falling asleep during the day from a needed nap. If you look up "sleep paralysis" on the Internet, you will see that it is often related to what people think are out-of-body experiences (OBEs), demonic possessions, and many other psychic phenomena.

I have had several controlled sleep-paralysis episodes now and find the state of mind to be extremely interesting, almost like the "extreme sport" version of sleep research! If you know what to expect, the hallucinations and pure weightlessness feeling can be quite a wild ride. I found that using the synchro brain machine or even the alpha meditation goggles set to the 6-Hz range while trying to sleep during a quiet afternoon was a good way to trigger a sleep-paralysis episode. Having no caffeine that day or other stimulants and being a bit on the tired side also were factors that improved the chances of reaching this rare and almost mystical state. Find a quiet, comfortable place to lie on your back, and just try to clear your mind while using the brain entrainment system. I must warn you, though, before you consider trying to purposely put yourself in a hypnagogic state, read as much as you can on sleep paralysis and lucid dreaming so that you know what you are about to encounter. Going through a session of sleep paralysis may be the strangest, scariest, and most bizarre state of mind you have ever encountered, so be prepared! Okay, enough talk—let's solder some wires now!

The synchro brain machine is a pair of independently tunable audio oscillators that feed their outputs into the right and left sides of a stereo headphone worn by the user. There are a few examples of commercial units available that have many nice features included, such as a digital frequency readout, multiple waveforms, and even the ability to add white noise or mix the binaural beat frequencies with some type of external audio source. This extremely "bare bones" version of the device actually will do most of this as well, although you will need to use your multimeter to see the digital frequency display and some external audio source such as an MP3 player to mix in your music or meditation media.

Figure 19-1 shows the synchro brain machine schematic, and the two oscillators made from the two 555 timers (IC1 and IC2) can be seen clearly. The third IC (IC3) is an LM358 op amp that is working as a simple comparator that will flash an LED each time the two audio frequencies cross each other or become synchronous for a short period. The resulting LED flash rate will be the same binaural beat that your brain will hear when presented with the two different audio frequencies. Each oscillator output is fed into the right and left sides of a stereo headphone jack and to an optional jack that will allow you to monitor the frequency of each oscillator channel. To change the two oscillator frequencies, you just turn the variable resistors (VR1 and VR2) to some frequency that is comfortable to listen to. It is the resulting binaural-beat frequency that counts, not the initial oscillator frequencies.

The breadboarded synchro brain machine circuit is shown in Figure 19-2, running on a pair of AA batteries. By using AA or even AAA batteries, you could build the entire unit into a headphone casing or some very small box that could be affixed directly to the headphones, making a very portable device. I preferred the external box version so that I could have the frequency adjustment controls near my hand while relaxed. When you first power up the unit, the indicator LED probably will be steady until you tune both variable resistors to the exact same spot. When your oscillators are only slightly out of phase, you will see the LED flash very slowly or several times per second. Listening to the output and tuning the controls as if you were tuning a guitar is another way to get the oscillators close to phase. As soon as you have them outputting similar frequencies, you will hear

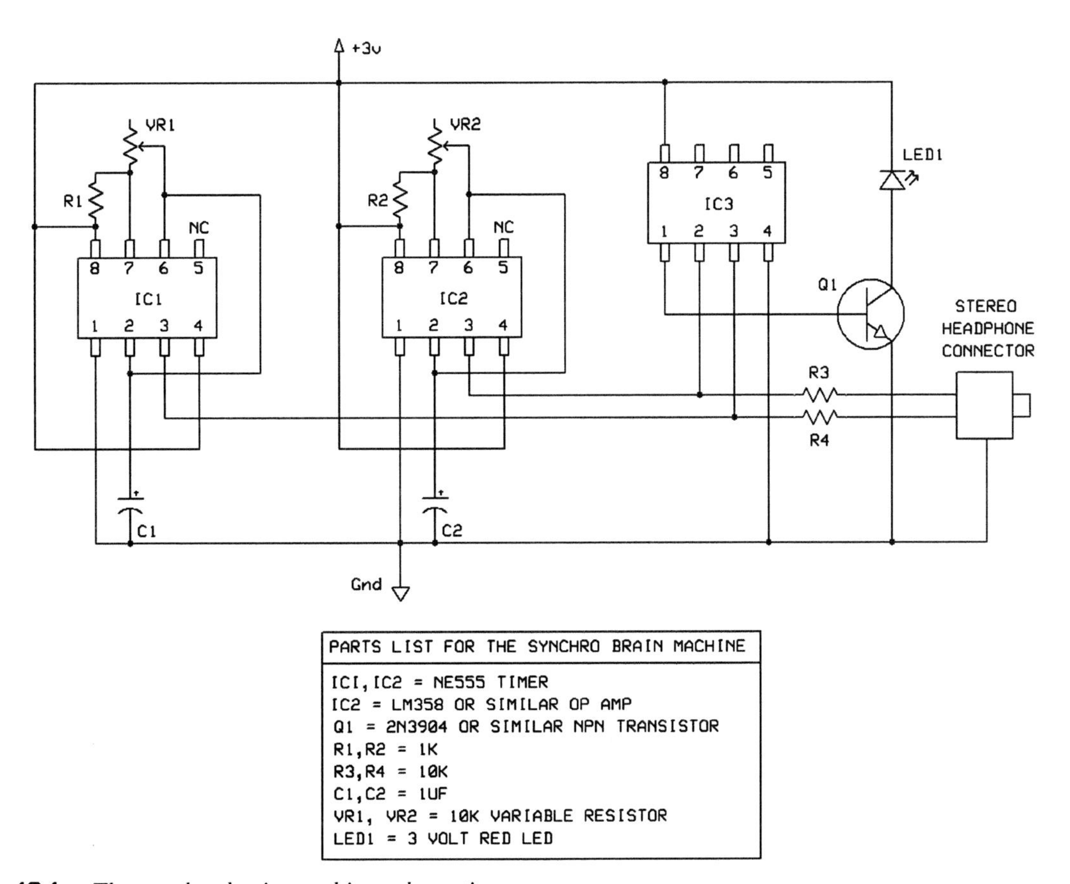

Figure 19-1 *The synchro brain machine schematic.*

Figure 19-2 *The synchro brain machine circuit.*

that "wong, wong, wong, wong" sound that I was referring to earlier. This is the binaural beat.

From my research into binaural beats, it seems that frequencies between 200 and 400 Hz are used most commonly because they are easy to listen to compared with the very high or low frequencies. There is plenty of range in the oscillators to go well beyond or below 300 Hz, so you can experiment with the frequencies you are most comfortable with. Figure 19-3 shows the readout on my basic multimeter, indicating that the left channel is set to 316 Hz. I have the right channel set to 306 Hz, which places the binaural-beat frequency at 10 Hz, right into the alpha-wave range.

If you built the circuit on a small enough piece of perf board and use a pair of AAA batteries or even a camera battery, you might be able to cram the works inside a large set of headphones, making a nice all-in-one device. I wanted to add another stereo jack to the unit so that I could easily attach my frequency counter or input an external audio source into the sound stream. Thus I opted for the conventional black-box enclosure shown in Figure 19-4. Having the remotely located frequency control also was convenient when I was relaxing so I did not have to move my arms to adjust the controls. I became so used to the alpha frequencies I so often used that I could set them without actually needing the digital readout. This is the same skill a musician uses when picking out a certain note or key.

Figure 19-5 shows my completed synchro brain machine—the true source of my magical powers and psychic abilities! If you find the output of the device to be too loud or too quiet, then change the values of R3 and R4 or replace them with a pair of 1K variable resistors so that you can set the volume to a level that you find most comfortable. If you plan to use the same basic frequencies most of the time, then you even could replace the frequency-control variable resistors with smaller potentiometers or have multiple preset banks of potentiometers and a select switch to instantly choose a "program" that you have preset. I preferred to simply listen to the

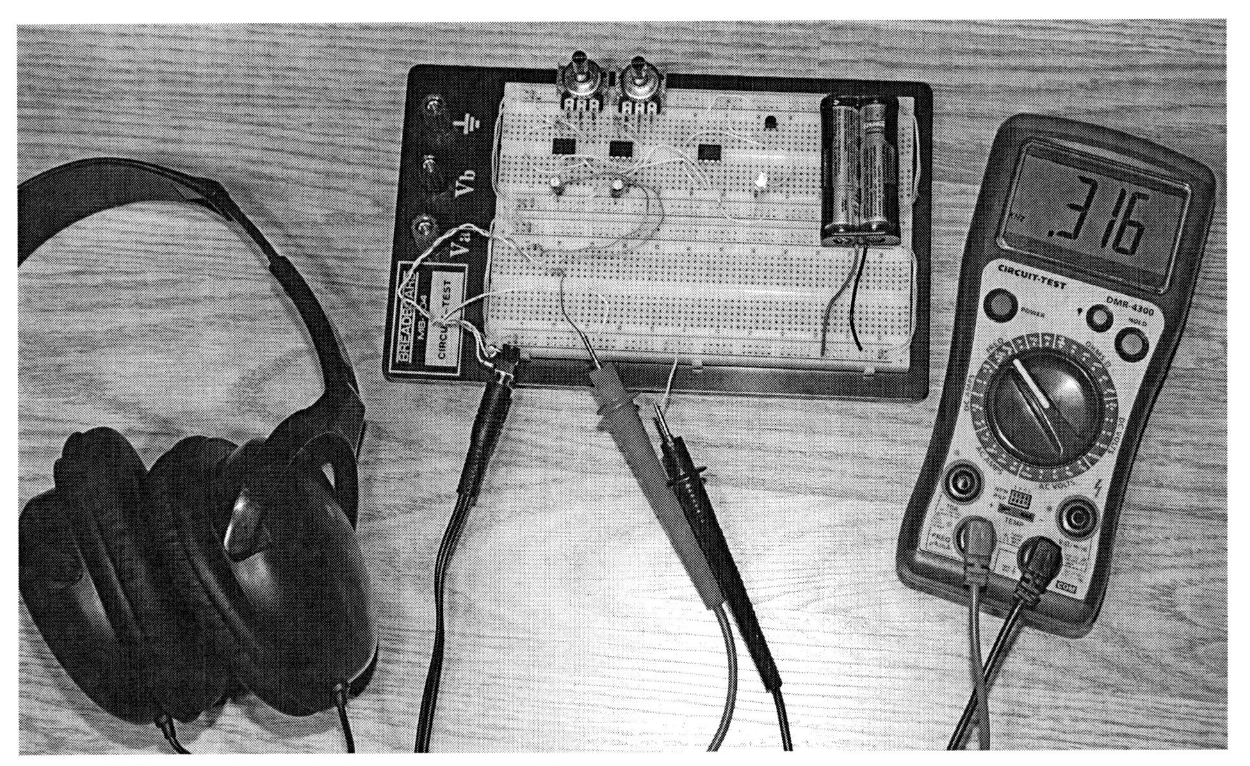

Figure 19-3 *Displaying the frequency on a multimeter.*

Figure 19-4 *Boxing up the synchro brain machine circuit.*

Figure 19-5 *The completed synchro brain machine.*

output and set the controls to a base frequency and beat frequency that felt right at the time. Some frequencies will just seem to feel right after you work with the device for some time. Maybe the brain learns to crave certain states much the way your mind craves certain tastes when the body requires refueling.

Before you start experimenting with this device, do a little Internet research on "binaural beats" and "brain entrainment" so that you will understand the way the brain responds to this technology and how you can use it to your benefit. As with all things in the public domain, you will have to toss away the far ends of the debate and look to the middle to learn anything useful. On one end of the scale, I have read that brain entrainment is the path to complete enlightenment and psychic powers, and on the end, I have read reports of brain entrainment being the most dangerous thing on earth. As a true scientist, I will leave you to make your own decisions based on actual research!

Some interesting modifications you might want to consider include the addition of a light stimulus such as the goggles presented in the alpha meditation section or the color-therapy device. How about adding white noise or mixing the binaural beats with some meditation or lucid-dream training media? There are so many ways you can enhance this unit, so enjoy the journey, but be prepared for the unexpected!

Conclusion

The electronics hobby is both entertaining and mind expanding, much like many of the devices presented in this book. With only a handful of inexpensive semiconductors and a breadboard, you can hack together just about any type of device in your "evil genius" workshop in a few evenings. Sure, there will be "bugs" and even the odd puff of smoke as you try out new designs, but as with all things, practice and patience will bring results.

Many of the projects presented here can be mixed and matched together to create entirely new devices, and there is always room for modifications and improvements, so feel free to let out your "crazy ideas" and try out new designs. When you want to further research a certain aspect of biofeedback or electronic design, the Internet is the ultimate source for free and comprehensive information, and there are many forums and user groups that you can become a part of.

When "plugging" yourself into any home-built device, remember to consider the safety aspects of your design, especially if there is a connection between the hardware and some other appliance, such as a computer or an external power supply. It is also a good idea to have a clean bill of health from your doctor before subjecting yourself to any type of sound- or light-pulsing device to reduce the risk of accidentally inducing health issues such as an epileptic seizure. Although the risk of harm is extremely remote when messing around with low-power projects running from a single battery, *safety first* is always a good rule to follow.

Have fun, and don't be afraid to take your ideas from the drawing board to the circuit board. You just never know what you might invent by thinking outside the box!

Appendix

Code Listings

Listing 3-1 Body temperature monitor source code

```
' DEFINE TARGET = MEGA88 @ 4MHZ
$regfile = "M88def.dat"
$crystal = 4000000

' CONFIGURE IO PORTS
Config Scl = Portb.2
Config Sda = Portb.1
Config Portd.0 = Output
Config Portd.1 = Output
Config Portd.2 = Output
Config Portd.3 = Output
Config Portd.4 = Output
Config Portd.6 = Output
Config Portd.7 = Output
Config Portb.3 = Output
Config Portb.4 = Output

' DEFINE VARIABLES
Dim Led(10) As Byte
Dim Msb As Byte
Dim Lsb As Byte
Dim A As Byte
Dim B As Byte
Dim C As Byte
```

```
' DEFINE LED DIGITS
Led(1) = 8
Led(2) = 187
Led(3) = 97
Led(4) = 49
Led(5) = 178
Led(6) = 52
Led(7) = 4
Led(8) = 185
Led(9) = 0
Led(10) = 48

' ********** MAIN LOOP **********
Main:
Do

' READ DS1621 TEMPERATURE
I2cstart
I2cwbyte &H90
I2cwbyte &HEE
I2cstop
I2cstart
I2cwbyte &H90
I2cwbyte &HAA
I2cstop
I2cstart
I2cwbyte &H91
I2crbyte Msb , Ack
I2crbyte Lsb , Nack
I2cstop

' DISPLAY DATA ON LEDS
A = Msb Mod 10
B = Msb \ 10
Gosub Ledshow

' RESTART MAIN LOOP
Loop
End

' ********** LED DISPLAY ROUTINE **********
Ledshow:
Portb.4 = 0
Portb.3 = 1
C = A + 1
Portd = Led(c)
```

```
Waitms 2
Portb.4 = 1
Portb.3 = 0
C = B + 1
Portd = Led(c)
Waitms 2
Return
```

Listing 5-1 The heart rate monitor Basic program

```
' DEFINE TARGET = MEGA88 @ 4MHZ
$regfile = "M88def.dat"
$crystal = 4000000

' CONFIGURE IO PORTS
Config Portd.0 = Output
Config Portd.1 = Output
Config Portd.2 = Output
Config Portd.3 = Output
Config Portd.4 = Output
Config Portd.6 = Output
Config Portd.7 = Output
Config Portb.3 = Output
Config Portb.4 = Output
Config Portb.5 = Output
Config Portb.2 = Output

' DEFINE VARIABLES
Dim Led(10) As Byte
Dim A As Byte
Dim B As Byte
Dim C As Byte
Dim D As Word
Dim E As Word
Dim F As Integer
Dim G As Word
Dim H As Word
Dim J As Integer
Dim K As Integer

Dim X As Byte

' DEFINE LED DIGITS
Led(1) = 8
```

```
Led(2) = 187
Led(3) = 97
Led(4) = 49
Led(5) = 178
Led(6) = 52
Led(7) = 4
Led(8) = 185
Led(9) = 0
Led(10) = 48

' SET DEFAULT HEART RATE
K = 60

' START ADC RUNNING
Start Adc

' ********** MAIN LOOP **********
Main:
Do

' READ ADC VALUE
D = Getadc(0)

' GET ADC CHANGE SINCE LAST
F = D - E
F = Abs(f)
E = D

' test bench 1 second = 164 clk
'X = X + 1
'F = 0
'If X = 200 Then
'X = 0
'F = 10
'End If

' HEART BEAT FILTER
If G > 0 Then G = G - 1
If F > 4 And G = 0 Then G = 40

' HEART BEAT LED FLASHER
If G = 1 Then Portb.2 = 0
If G = 20 Then Portb.2 = 1

' CALCULATE HEART RATE PER MINUTE
H = H + 1
```

```
If G = 1 Then
J = 9840 / H
H = 0
End If

' SLOWLY ADJUST RUNNING AVERAGE
If G = 1 Then
If K > J Then K = K - 1
If K < J Then K = K + 1
End If

' DISPLAY DATA ON LEDS
If K > 99 Then
C = K Mod 10
B = K \ 10
B = B Mod 10
A = K \ 100
End If
If K < 100 Then
C = K Mod 10
B = K \ 10
A = 0
End If
If K < 10 Then
C = K
B = 0
A = 0
End If
Gosub Ledshow

' RESTART MAIN LOOP
Loop
End

' ********** LED DISPLAY ROUTINE **********
Ledshow:
Portb.5 = 1
Portb.4 = 0
Portb.3 = 0
D = A + 1
Portd = Led(d)
Waitms 2
Portb.5 = 0
Portb.4 = 1
Portb.3 = 0
D = B + 1
```

```
Portd = Led(d)
Waitms 2
Portb.5 = 0
Portb.4 = 0
Portb.3 = 1
D = C + 1
Portd = Led(d)
Waitms 2
Return
```

Listing 13-1 Optical sensor-based program code

```
' ********** DREAM MASK PHOTOTRANSISTOR VERSION **********
' DEFINE TARGET = MEGA88 @ 4MHZ
$regfile = "M88def.dat"
$crystal = 4000000

' CONFIGURE IO PORTS
Config Portb.0 = Output
Config Portb.1 = Output
Config Portb.2 = Input
Config Portb.3 = Output

' DEFINE VARIABLES
Dim A As Word
Dim B As Word
Dim C As Integer
Dim D As Byte
Dim E As Byte

' START ADC RUNNING
Start Adc

' ********** MAIN LOOP **********
Main:
Do

' READ ADC VALUE
A = Getadc(5)

' GET ADC CHANGE SINCE LAST
C = A - B
C = Abs(c)
```

```
' DETECT EYELID MOVEMENT
If C > 2 Then

' FLASH TEST LED
Portb.0 = 1
Waitms 200
Portb.0 = 0
Waitms 200

' COUNT EYELID REM SIGNALS
D = D + 1

' FLASH GOGGLE LED 100 TIMES AFTER 20 REM SIGNALS
If D = 20 Then
Portb.3 = 1
For B = 1 To 100
Portb.1 = 1
Gosub Button
Waitms 400
Portb.1 = 0
Gosub Button
Waitms 400
Next

' RESET AND START OVER
Portb.3 = 0
D = 0
End If
End If

' RESET LAST ADC VALUE
B = A

' CHECK FOR BUTTON PRESS
Gosub Button

Loop
End

' ********** BUTTON PRESS **********
Button:
If Pinb.2 = 0 Then Return

' RESET SYSTEM
D = 0
B = 100
```

```
E = 0

' ENTER TEST MODE
While Pinb.2 = 1
Waitms 20
E = E + 1
If E = 200 Then
E = 0
Portb.0 = 1
Goto Setup
End If
Wend

Return

' ********** SETUP MODE **********
Setup:
Do

' GET ADC VALUE
A = Getadc(5)
C = A - B
C = Abs(c)
If C > 2 Then

' SETUP MODE LED FLASH
Portb.1 = 1
Waitms 50
Portb.1 = 0
Waitms 100
End If
B = A

' EXIT SETUP MODE
While Pinb.2 = 1
Waitms 10
E = E + 1
If E = 200 Then
Portb.0 = 0
E = 0
Goto Main
End If
Wend

Loop
```

Listing 13-2 Accelerometer sensor–based program code

```
' ********** DREAM MASK ADXL202 VERSION **********
' DEFINE TARGET = MEGA88 @ 4MHZ
$regfile = "M88def.dat"
$crystal = 4000000

' CONFIGURE IO PORTS
Config Portb.0 = Output
Config Portb.1 = Output
Config Portb.2 = Input
Config Portb.3 = Output

' DEFINE VARIABLES
Dim A As Integer
Dim B As Integer
Dim C As Integer
Dim D As Byte
Dim E As Byte

' ********** MAIN LOOP **********
Main:
Do

' READ ADXL202 VALUE
Pulsein A , Pinc , 5 , 0

' GET ADXL CHANGE SINCE LAST
C = A - B
C = Abs(c)

' DETECT EYELID MOVEMENT
If C > 2 Then

' FLASH TEST LED
Portb.0 = 1
Waitms 200
Portb.0 = 0
Waitms 200

' COUNT EYELID REM SIGNALS
D = D + 1

' FLASH GOGGLE LED 100 TIMES AFTER 20 REM SIGNALS
```

```
If D = 20 Then
Portb.3 = 1
For B = 1 To 100
Portb.1 = 1
Gosub Button
Waitms 400
Portb.1 = 0
Gosub Button
Waitms 400
Next

' RESET AND START OVER
Portb.3 = 0
D = 0
End If
End If

' RESET LAST ADXL VALUE
B = A

' CHECK FOR BUTTON PRESS
Gosub Button

Loop
End

' ********** BUTTON PRESS **********
Button:
If Pinb.2 = 0 Then Return

' RESET SYSTEM
D = 0
B = 100
E = 0

' ENTER TEST MODE
While Pinb.2 = 1
Waitms 20
E = E + 1
If E = 200 Then
E = 0
Portb.0 = 1
Goto Setup
End If
Wend
```

```
Return

' ********** SETUP MODE **********
Setup:
Do

' GET ADXL VALUE
Pulsein A , Pinc , 5 , 0
C = A - B
C = Abs(c)
If C > 2 Then

' SETUP MODE LED FLASH
Portb.1 = 1
Waitms 50
Portb.1 = 0
Waitms 100
End If
B = A

' EXIT SETUP MODE
While Pinb.2 = 1
Waitms 10
E = E + 1
If E = 200 Then
Portb.0 = 0
E = 0
Goto Main
End If
Wend

Loop
```

Listing 16-1 The clairvoyance tester Basic source code

```
' DEFINE TARGET = MEGA88 @ 4MHZ
$regfile = "M88def.dat"
$crystal = 4000000

' CONFIGURE IO PORTS
Config Portd.0 = Output
Config Portd.1 = Output
Config Portd.2 = Output
Config Portd.3 = Output
```

```
Config Portd.4 = Output
Config Portd.6 = Output
Config Portd.7 = Output
Config Portb.3 = Output
Config Portb.4 = Output
Config Portb.5 = Output
Config Portb.1 = Input
Config Portb.2 = Input

' DEFINE VARIABLES
Dim ___rseed As Word
Dim Led(10) As Byte
Dim A As Byte
Dim B As Byte
Dim C As Byte
Dim D As Byte
Dim E As Byte
Dim F As Byte

' DEFINE LED DIGITS
Led(1) = 8
Led(2) = 187
Led(3) = 97
Led(4) = 49
Led(5) = 178
Led(6) = 52
Led(7) = 4
Led(8) = 185
Led(9) = 0
Led(10) = 48

' ********** MAIN LOOP **********
Main:
Do

' PRESSED RANDOMIZE BUTTON
If Pinb.1 = 0 Then
Gosub Roller
End If

' CHANGE GUESS BUTTON
If Pinb.2 = 0 Then
A = A + 1
If A = 10 Then A = 0
While Pinb.2 = 0
Gosub Ledshow
```

```
Wend
End If

' RANDOM SEED COUNTER
___rseed = ___rseed + 1

' DISPLAY DATA ON LEDS
Gosub Ledshow

' RESTART MAIN LOOP
Loop
End

' ********** LED DISPLAY ROUTINE **********
Ledshow:
Portb.5 = 1
Portb.4 = 0
Portb.3 = 0
D = A + 1
Portd = Led(d)
Waitms 2
Portb.5 = 0
Portb.4 = 1
Portb.3 = 0
D = B + 1
Portd = Led(d)
Waitms 2
Portb.5 = 0
Portb.4 = 0
Portb.3 = 1
D = C + 1
Portd = Led(d)
Waitms 2
Return

' ********** RANDOM ROLL ROUTINE **********
Roller:

' RANDOMIZE RESULT
For E = 1 To 50
B = Rnd(10)
For F = 1 To 10
Gosub Ledshow
Next
Next
```

```
' ADJUST SCORE
If A = B Then
C = C + 1
Else
C = C - 1
End If
If C = 255 Then C = 0
Return
```

Index

CPSIA information can be obtained
at www.ICGtesting.com
Printed in the USA
FFOW01n2021021115
18247FF